COLLINS GEM GUIDE

THE

FARM

Text by Jon Newton and James France
Illustrations by Tony Lodge

HarperCollins*Publishers*

HarperCollins Publishers
PO Box, Glasgow, G4 0NB

First published   1983

Reprint 10 9 8 7

© Jon Newton and James France (text) 1983
   Tony Lodge (illustrations) 1983

ISBN 0 00 458813 4

Printed in Great Britain by
HarperCollins Manufacturing, Glasgow

Contents

About This Book

The emphasis of this book is on British agriculture and its diversity, although there is coverage of other European countries. Eleven sections, each largely self-contained, cover livestock, poultry and crops. The twin aims of the book are to enable the reader to identify the more common livestock breeds and plant species used in agriculture, and to understand what the farmer is doing and why. As well as aiding identification, therefore, the book sets out to describe the important practices of farming, such as animal feeding, crossbreeding, ploughing and harvesting, and the end-products of the farmer's labours. Inevitably, a certain amount of biology has been included, but this has been kept to a minimum.

Finally, there is much to be seen and enjoyed on our farmlands. This land is, however, the farmer's livelihood so when walking across it help him by observing the **Country Code**. In particular, guard against all risk of fire and leave no litter. Keep to paths across farmland, fasten all gates, and do not damage fences, hedges and walls. At all times, protect and respect wildlife, wild plants and trees.

Introduction

Farming is the major activity in the countryside and is second only to hunting as an ancient pursuit of man. This book is for people who want to know more about what happens on the farm. It is for those not satisfied with answers such as that is a cow, a sheep, corn or grass, but who want to know more: what breed of cow or type of corn and why? There is infinite diversity in agriculture and a great deal of skill based on tradition, education and science, and nowhere is this diversity more apparent than in Britain and Europe.

European Agriculture

Climate, terrain and soil-type all affect the use the land can be put to. For example, most of Finland and Sweden is afforested, while there are very few forests in Ireland and none at all in Iceland. Two-thirds of Denmark is under arable cultivation, whereas arable land accounts for only 2 per cent of Norway and less than 1 per cent of Iceland. Most of Ireland is grassland. Britain and Switzerland, too, have high proportions of grassland.

The percentage of the population engaged in agriculture varies widely across Europe. Britain and Belgium, where about one in twenty work the land, are at the lower extreme. At the other end of the spectrum are Spain, Portugal and Greece, where more than two-fifths of the population are involved in farming.

Cattle are the most important of European livestock. They are kept primarily for meat and milk though are still used as draught animals in some parts. France has over 20 million head of cattle and West Germany, the United Kingdom and Italy all have large national herds. Dairy products are a major export for countries such as Denmark, France, Ireland and Holland.

There are also a large number of sheep, kept mainly for meat and wool though sometimes for milk. Britain, with a national flock of over 30 million, has the most, followed by Spain and France, each with just under half this number. Some European countries have more sheep than people; Iceland, for example, has four sheep

per person. In continental Europe some flocks are still nomadic – for instance in parts of France, where sheep move up to high alpine pastures for the summer, and in intensively farmed Holland, where Texel sheep graze freely along the dykes.

Pigs are important in much of Europe. Pig farming and potato growing often go together. West Germany has the largest national breeding herd, with over 22 million animals, and their offspring constitute two-thirds of the country's meat production. The most common name among European pig breeds is the Landrace, or 'local breed'.

Of less importance nowadays are goats and horses. Most of Europe's goats are found in Greece, with over 4 million, France, with just over a million, and Italy, with nearly a million. Numbers in Switzerland, the original home of most European goat breeds, have dropped markedly in recent years as grazing goats in forested

areas is now prohibited. The use of the working heavy horse has declined considerably in Europe, due to the tractor, though France still has more than a million of them. Mules and donkeys are commonly used for transport in some Mediterranean areas – Greece for example has 250 000 donkeys and 100 000 mules. They are also popular in Spain and Portugal.

As well as livestock, vast numbers of poultry are to be found. These are mostly hens kept for eggs and for the table. France and Italy lead the way with national flocks of 178 and 121 million birds respectively.

Grasslands play a vital role in feeding ruminant livestock such as cattle and sheep. As well as summer grazings, they provide grass for conserving as hay and silage for winter feeding. The major grassland countries are the moist, temperate ones, such as Britain and Ireland. Some Scandinavian countries make considerable use of clover for hay, and lucerne is of increasing importance in France and Spain.

Cereal crops are of great significance to

European agriculture. They provide grain for making basic human foods such as bread, and corn for feeding to animals and poultry. Wheat is the most important, with France producing the greatest tonnage, followed by Italy and West Germany. Flour milling is one of the major industries of northern France and the big Italian pasta industry is based on the soft grain grown in the Po Valley. Barley, being hardier, is more important than wheat in more northerly countries. Oats is the commonly grown cereal in Finland and Sweden, and rye is grown widely in West Germany on the sandy soils. Maize, too, is a major crop in warmer countries such as France, Italy, Portugal and Spain.

Other arable crops include roots, such as potatoes and sugar beet, peas, beans, oilseed rape, flax, cabbage and kale. Over half the arable area of West Germany is used for growing potatoes. Sugar beet is a major crop in France, Italy and Britain, and flax, used for making linen, is widely grown in Belgium and Holland.

Cattle

Cattle belong to the genus *Bos* (Latin ox), which includes such species as Bison, Buffalo and Yak, as well as domestic cattle. All are believed to have evolved from a common ancestor that lived in Asia during the Pliocene period, 2–7 million years ago. One of the first distinct species to have developed in the Pliocene was *Bos primigenius*, the remains of which have been found in Egypt, Europe and India. These animals were the forerunners of the domestic cattle we know today. In Europe *Bos primigenius* took the form of the now extinct wild Aurochs, the last of which was killed in the forests of Poland in 1627. However, by crossing existing breeds of cattle the Aurochs has been 'reconstructed', and a semi-wild herd now lives in Poland.

The domestication of cattle began in India and the Middle East between 6000 and 4000 BC around the valleys of the Indus, the Tigris and Euphrates, and the Nile. Domestication spread as people from the Middle East migrated south and west through Africa and then north and west into Europe. There were no domestic cattle either in the Americas or in Australia until European settlers introduced them.

Cattle are kept primarily for work, meat and milk. Some breeds fulfil all three functions and are thus triple purpose. However, the number of domestic cattle used for work is decreasing rapidly, though it is still possible to see draught oxen at work in parts of Europe. Chianina cattle, for example, can occasionally be seen pulling a cart or plough in northern Italy. In Britain, the draught oxen had virtually disappeared from the agricultural scene by the onset of the First World War.

Chianina cattle in northern Italy

Meuse-Rhine-Ijssel A popular
breed from Holland

Nowadays most cattle breeds are classified as either dairy, beef or dual purpose
(meat and milk). A typical dairy cow, of
which the Ayrshire is a good example, has a
thin, wedge-shaped body, with little flesh
and prominent bones. A typical beef type,
such as the Beef Shorthorn, has a solid,
rounded body, well-covered with flesh. A
dual-purpose animal, such as the
Meuse–Rhine–Ijssel from Holland, is an
intermediate type, having the capacity to
produce a good supply of milk and beef of an
acceptable quality.

Ayrshire
A hardy, healthy breed capable of producing high yields of good milk

Beef Shorthorn
Used for crossing with dairy breeds to produce beef calves

13

Points of a cow Each part of a bull or cow has a particular name which is used by cattlemen in describing their beasts. These 'points of conformation' are shown below. There are basic conformation characteristics which all cattle breeders concentrate on improving. With dairy cows, a well-developed udder is very important. It should be attached to the body of the cow with strong muscles and the teats must be evenly spaced, of medium size, and point straight downwards. With beef cattle the body should be long, wide and deep, and well-muscled, particularly in the hindquarters and back.

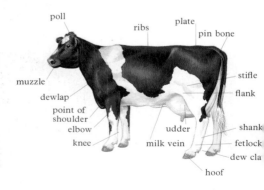

Feeding Cattle can obtain their nutritional requirements from grass, either fresh or in the form of hay or silage. They can also eat other products such as straw, cereals, swedes, kale, various industrial by-products, and concentrates – although the last, being expensive, are normally only fed to highly productive cattle such as lactating dairy cows and fast-growing beef cattle. In Britain, cattle are normally at pasture from spring to autumn, but in winter they are usually housed and stall-fed or fed in the yard. A mature cow eats 10–20 kg of dry matter a day and consumes even more water. A dairy cow, when not in milk, requires 30–40 litres of water a day, while a lactating cow might need more than 100 litres a day.

Friesians at a
silage face

15

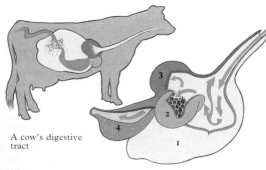

A cow's digestive tract

Digestion Ruminants have a large, complex stomach that enables them to eat forages, such as grass and straw, which contain large amounts of cellulose. It consists of four compartments: the rumen (or paunch) (1), reticulum (2), omasum (3) and abomasum (or true stomach) (4). Cellulose is broken down in the rumen by millions of micro-organisms which reduce it to simple compounds that the animal can then digest. Digestion beyond the rumen is similar to that of simple-stomached animals. As little chewing is done before food is first swallowed to the rumen, cattle spend a considerable part of their day regurgitating and rechewing it. This is called ruminating or 'chewing the cud'.

16

Health and diseases The major diseases that affect cattle are infections caused by bacteria and viruses. Bacterial infections include mastitis and brucellosis while viral ones include foot and mouth and viral pneumonia. Parasites are a frequent cause of disease. External parasites such as ringworms and ticks usually feed on the animal's skin or hair, while internal ones such as liver flukes and lungworms (husk) live inside it. Other disorders include milk fever and bloat, which is caused by an excessive build-up of gas in the rumen. Signs of a healthy cow include a moist nose, bright eyes, a good appetite, and an alert and lively posture.

Using a trocar and cannula to relieve bloat. The rumen must be punctured in order to release excess gas

The science of breeding The science of breeding is known as genetics. The basic laws of genetics were discovered by an Austrian monk, Gregor Mendel, in the mid-nineteenth century. Mendel worked with peas, but the principles he discovered apply equally to animals. His experiments showed that inherited characteristics are passed on by factors called *genes*. Cattle carry a large number of different genes which affect every aspect of their appearance and function. Some of these characteristics we can see, such as colour of coat, pattern of markings, and type of horns; others we cannot see, such as milk yield, growth rate and hardiness.

Genes are carried in rod- or V-like structures called *chromosomes*, which are found in the nucleus of all the cells of the body. Each cell has the same complement of chromosomes, and hence of genes. Chromosomes are always present in pairs, the two members of a pair being identical in shape and size. One member of each pair comes from each parent: thus genes are also paired.

Consider the inheritance of colour of coat in Shorthorn cattle (pictured opposite). If a pure white Shorthorn bull is mated with a pure red Shorthorn cow (1), all their calves will be roan (2). The bull carries a pair of genes which produce white hairs and the cow a pair producing red hairs. The calves

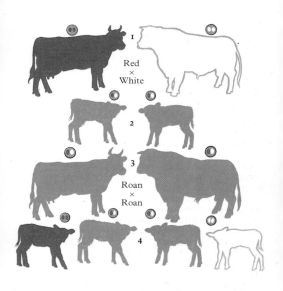

Red
×
White

Roan
×
Roan

receive one gene from each parent and thus develop a mixture of red and white hairs. The roan animal will subsequently be able to pass on to its calves either a red or a white gene, but not both. If two roan animals are mated (3), their calves will receive either a red gene from both parents, a white gene from both parents, a red gene from father and a white gene from mother, or a white gene from father and a red gene from mother (4). The calf receiving two red genes will be red; the calf receiving two white genes will be white; and the calf receiving one red gene and one white gene will be roan like its parents. The chances of a red calf are therefore 1 in 4, of a white calf 1 in 4, and of a roan calf 2 in 4.

When a cow and a bull of different colours are mated, their calves are not necessarily of an intermediate colouring (pictured opposite). For example, if the black Aberdeen Angus is crossed with a Red Poll (1), the resulting calves are always black (2). This is because, for many gene pairs, one gene is *dominant* and the other *recessive*. The Aberdeen Angus × Red Poll calf really contains one black gene and one red gene, but the black dominates the red and so the calf appears black. If two Aberdeen Angus × Red Poll calves are mated (3) their calf will receive either a red gene from both parents, a black gene from both parents, one red from its father and one black from its mother, or one black gene from its father and one red gene from its mother (4). As the black dominates the red gene, the chances of a red calf are 1 in 4 and a black calf 3 in 4. Some common dominant characteristics in cattle are black coat to red coat; a single-

coloured coat to a pied coat; a polled head to a horned head; and a white Hereford face to other types of face. Therefore, when a Hereford bull is mated with a British Friesian cow, we get the familiar Hereford × Friesian calf with its black, single-coloured coat and white Hereford type face.

21

Mating, pregnancy and calving Cows are served either naturally by the bull or, more commonly, by artificial insemination (AI). With AI, semen is taken from bulls kept at AI centres and then frozen and stored. It can be kept for some considerable time in this way. The semen is subsequently used on farms by trained inseminators who insert it into the cows. A cow can only be served when she is 'on heat', a period termed oestrus or bulling, which occurs about every 21 days and lasts for about a day. The duration of pregnancy is on average 283 days.

Just before calving the cow becomes uneasy and often seeks a secluded spot in which to give birth. The head and feet of the emergent calf should appear first; the cow will need some help if the calf is not in this position. After the calf is born the afterbirth (or cleansings) come away from the mother. The cow cleans her new-born calf by licking it.

The first milk taken from the cow after calving is called colostrum (or beestings), which is dark in colour and differs from normal milk in composition. It is vital that the calf receives colostrum within a few hours of being born, for it contains important antibodies which protect the calf against disease. The cow's milk returns to normal within three or four days.

Calf foetus at one month (left) and nine months

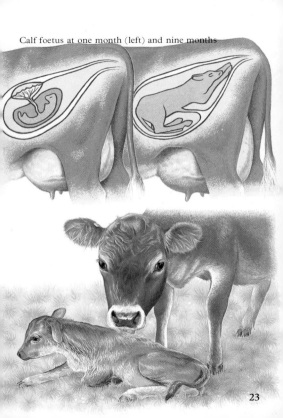

23

Dairy breeds Fifty years ago, most of Britain's milk came from dual-purpose breeds such as the Dairy Shorthorn. Nowadays our milk comes from specialist dairy

British Friesian
A large, high-yielding dairy cow capable of producing good beef calves

Jersey
A small Channel Island breed, with a typical dairy conformation, noted for its very rich milk

24

breeds, the most important being the British Friesian which accounts for some 80 per cent of the dairy cows in England and Wales. Other important dairy breeds are the Ayrshire, the Guernsey and the Jersey.

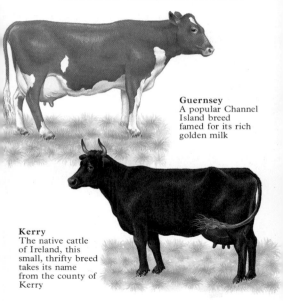

Guernsey
A popular Channel Island breed famed for its rich golden milk

Kerry
The native cattle of Ireland, this small, thrifty breed takes its name from the county of Kerry

Finncattle
A hardy breed found in Finland,
giving good yields of rich milk

Norwegian Red
A productive dairy breed common to
Norway

Housing Dairy cows are generally housed during the winter. In the past they were housed in cowsheds and tied up individually in standings, but today loose-housing is practised on most dairy farms. There are two methods: one using covered, strawed yards and the other using cubicles or kennels. The first way is a simple, flexible system, but requires a lot of straw. The second combines the advantages of the straw yard and the cowshed – the cows can lie in their stalls or wander about the exercise area. Cubicles are installed in covered yards or existing buildings, while with cow kennels (pictured below) the stall divisions form part of the structure carrying the roof and walls.

Artificially reared
calves must first be
taught to feed
from a container

Most calves are
dehorned, with a
hot iron and
anaesthetic, at 5–6
weeks. Males,
unless breeding
stock, are also
usually castrated

28

Calf rearing In Britain most calves are now taken from their mothers at birth and reared artificially. The calf is given colostrum for the first three or four days of its life and after that is fed a milk substitute for five to six weeks, together with a special concentrate mixture and good quality hay. It is then weaned and fed only on solids. The young calf cannot deal with much bulky food as it is born with only a simple (or true) stomach – the rumen and the other compartments develop later.

The calves are kept in a calf house in individual pens for the first month or so to prevent them suckling one another and to reduce the risk of disease. Later, calves of the same size may be put together in groups.

Milking parlours Cows were once milked by hand over a bucket, which was then emptied into a churn. They are now milked in specially designed and labour-saving milking parlours where the cow moves into one of a limited number of spaces near the herdsman to be milked and then walks out again afterwards. In early milking parlours the milk was piped directly from the cow to a churn by way of a cooler. The heavy churns then had to be moved to a stand near the farm entrance where a lorry would collect them. Nowadays, the milk is passed from the parlour through a pipeline to a bulk storage tank where it is cooled by refrigeration. The milk is then emptied directly into a road tanker for collection.

Milking parlours are now of four basic designs. The herringbone parlour (pictured opposite), where the cow stands at an angle of 45° to the milker, is the most common for large herds. In the abreast parlour the cows stand side by side with their backs to the milker, and in the tandem parlour they stand in line with their sides to the milker. The rotary parlour, where the cows stand on a rotating platform with the milkers in the middle, is the most expensive and complex of the four systems.

Cows are not only milked in the parlour, but are often fed a ration of concentrates and are washed and cleaned.

Herring-bone milking
parlour, the most
popular for large herds

31

Milking Milk is produced continuously in the cow's udder in the months following calving. The udder therefore has to hold all the milk produced between milkings, which can exceed 15 litres in a high-yielding cow. The udder comprises four separate mammary glands or quarters. The upper part of each quarter consists of a mass of spongy tissue made up of millions of milk-producing cells grouped into microscopic hollow spheres called alveoli (1). Each alveolus is connected by small blood vessels to the blood supply from which it draws the necessary materials to synthesize milk. The cells secrete milk into the hollow centre of the alveolus and the milk leaves through a tiny duct. The ducts from the alveoli then lead into larger ducts that transport the milk to the udder cistern (2) located in the lower part of the quarter. Milk can be drawn off from the udder cistern through the teat, which contains a hollow cavity called the teat sinus (3). Milk leaves the teat sinus through a channel called the streak canal (4) which is protected from the outside by a sphincter muscle at the tip of the teat.

Most of the milk synthesized between milkings is stored in the alveoli and the small ducts of the upper udder; only a little milk passes down into the udder cisterns. Milking therefore consists of two phases. The first phase is the *let-down* of milk from

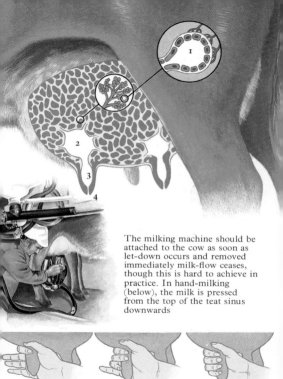

The milking machine should be attached to the cow as soon as let-down occurs and removed immediately milk-flow ceases, though this is hard to achieve in practice. In hand-milking (below), the milk is pressed from the top of the teat sinus downwards

the upper udder to the udder cisterns, and the second phase is its removal from the udder cisterns via the teats. Any nervous stress will make the cow hold up her milk, but let-down can be encouraged by several stimuli. With some cows milk flow begins as soon as they enter the milking parlour; others require feeding or having their udder washed in warm water. The removal of milk from the udder cisterns is carried out by hand or, much more commonly, by machine (both are pictured on p. 33). In hand milking, the milk is pressed from the top of the teat sinus downwards, so that the pressure on the milk above the streak canal is greater than the atmospheric pressure outside the teat, causing the sphincter muscle at the tip to open and milk to pass out. In machine milking, the difference in pressure is brought about by reducing the pressure on the outside of the tip of the teat at the centre of the teat cup. The first milk taken from the udder at milking should be examined for signs of mastitis or other problems.

Most dairy cows are milked twice a day. It is important to have a similar time interval between each milking and to keep to regular times. Hygienic milk production, with clean equipment and good milking methods, is essential if the modern dairy farmer is to be successful.

Composition Fresh milk provides a better balance of nutrients than possibly any other single food, although surprisingly it is 87 per cent water. Its composition varies with the type of cow, the feeding systems used, and other factors. Milk sold in Britain must contain at least 3 per cent butterfat (the amount of fat in the milk) and 6.5 per cent solids-not-fat (SNF).

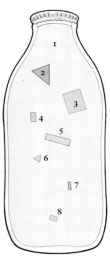

The constituents of milk and the proportions in which they are present in an average sample:

1 Water 87.51%
2 Fat 3.8%
3 Lactose 4.7%
4 Vitamins and minerals 0.76%
5 Casein 2.63%
6 Albumen 0.3%
7 Globulin 0.1%
8 Non-protein nitrogen 0.2%

Thus total solids present (2–8) amount to 12.49% and solids-not-fat (3–8) to 8.69%.
Casein, albumen and globulin are all proteins

Milk for liquid consumption In Britain we drink on average about 5 pints (2.7 litres) of liquid milk per person per week. Nearly all of it is heat-treated in one of three ways. The most common process is pasteurization, in which the milk is heated to 72°C for 15 seconds before being cooled to 10°C or below. Most of the bacteria are destroyed, without affecting colour and taste. Sterilized milk is heated to 110–115°C for half an hour in an airtight container, while ultra-heat treated (UHT) milk is flow-heated to 132°C for one second and then cooled rapidly. In both processes all the bacteria are destroyed, but with sterilization colour and taste are somewhat affected. Pasteurized milk will last for 2–3 days, sterilized milk for a week, and UHT milk virtually indefinitely.

Cultured, concentrated and dried milk Cultured milks, such as yoghurt, are prepared by souring fresh milk with bacteria that convert lactose to lactic acid, which preserves the milk solids in an edible form. Concentrated milk is produced by boiling milk at a reduced temperature (44–66°C) in a vacuum and drawing off surplus water vapour. Dried milk is made by spraying concentrated milk into a hot air chamber or spreading it on heated rollers to extract virtually all the remaining water.

Parmesan

Stilton

Cheese The basis of cheese-making is the addition of rennet to milk, which then separates into curds and whey. The curd contains the casein, most of the fat and some of the minerals, while whey is used as a by-product to feed cattle or in certain food-stuffs. Rennet consists of the two enzymes rennin and pepsin – rennin precipitates the curd and pepsin turns the curd into cheese.

Cheese is matured for various lengths of time, during which the micro-organisms and moulds develop that give cheese its flavour. Every area has its traditional method of cheese-making which is reflected in the great variety available, from the hard Parmesan, Stilton and Edam to the soft Brie, Camembert and Cottage Cheese. Cheese is also made from sheep's and goat's milk (see pp. 106 and 121).

Butter and cream Milk is divided into cream and separated milk by centrifugal force. Cream comprises the fat of the milk and a small amount of the other solids. The separated milk is made into skim milk powder. Cream is made into butter by being churned at speed until the fat globules coalesce. In the past this was done by revolving a wooden churn by hand, but today the process is fully mechanized. After churning, the liquid is drained off from the butter, which is then washed and sometimes salted.

dam

Brie

Beef breeds Pure beef breeds numerically account for only a minority of the British cattle population, but they are of great importance. Bulls from these breeds, such

Hereford
Now one of the dominant beef breeds in the world

Galloway
Polled, very hardy, and popular for crossing with other breeds to produce suckler cows

as the Hereford and Charolais, are widely used for crossing with our Friesian dairy herd to produce calves for commercial beef production.

Aberdeen Angus
A polled breed famous for its high-quality beef

Longhorn
This docile breed was once the most popular in Britain, but is now rare

Devon
This ruby-red breed has been prominent in the West Country for several centuries

Charolais
Popular French breed noted for its size, growth-rate and lean meat

Limousin
A large, fast-growing breed from west-central France

Blonde d'Aquitaine
A large French breed formed by amalgamating three similar strains of cattle

41

Beef production In Britain, most beef comes from calves produced by Friesian dairy cows – either purebred or, more often, crossbred with a beef bull such as the Hereford. A good beef carcass has plenty of lean, tender meat and a good proportion of high-priced cuts. The various cuts are shown opposite.

There are three basic systems of beef production: intensive, semi-intensive and traditional. Most beef calves are reared artificially and go into intensive or semi-intensive systems. Intensive systems produce lean beef cattle of no more than 12 months of age. The animals are housed, carefully managed and fed large amounts of concentrates. Semi-intensive systems also produce lean beef animals but at a heavier weight and over a longer period, usually 15–18 months. These animals are out at grass during the summer and are housed and yarded the rest of the time. Traditionally, beef was produced in Britain from store cattle. These are born and reared on one farm, usually in the grassland districts of the north and west, and then sold at up to two years old for finishing and fattening on grass or in yards elsewhere in the country.

There is little veal production in Britain, as only a small amount is eaten here. Veal is the meat of three-month old calves specially reared and fed.

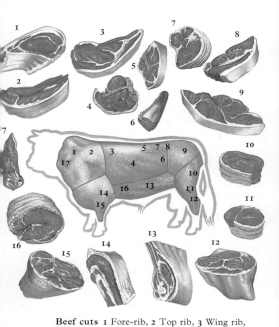

Beef cuts 1 Fore-rib, 2 Top rib, 3 Wing rib,
4 T-bone, 5 Porterhouse, 6 Fillet, 7 Sirloin,
8 Entrecote, 9 Rump, 10 Topside,
11 Silverside, 12 Leg, 13 Flank, 14 Brisket,
15 Shin, 16 Rolled brisket, 17 Blade bone

Dual-purpose breeds This century has seen marked changes in Britain in the relative importance of different breeds and types of cattle. Until recent times, dual-

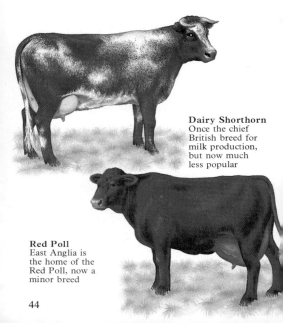

Dairy Shorthorn
Once the chief British breed for milk production, but now much less popular

Red Poll
East Anglia is the home of the Red Poll, now a minor breed

purpose breeds, producing both meat and milk, dominated our cattle population. They are now of little importance in Britain, but remain very important on the Continent.

Normandy
Its milk is used to make Camembert and other cheese

Simmental
A Swiss breed, and the most common dual-purpose one in Europe

45

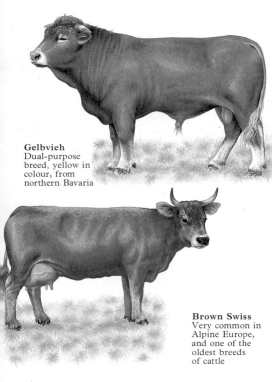

Gelbvieh
Dual-purpose breed, yellow in colour, from northern Bavaria

Brown Swiss
Very common in Alpine Europe, and one of the oldest breeds of cattle

By-products When cattle are slaughtered for meat they provide several useful by-products. The hide (1) is made into leather for shoes and other goods. The horns (2) are usually ground into fertilizer and the hooves (3) are used for buttons, fertilizer and neatsfoot oil. The bones (4) are also ground into fertilizer, provide glycerine for explosives and are boiled for glue, while blood is used as fertilizer and to make foam, glue and proteins. Inedible offal (5) is rendered down into tallow. Surgical ligatures are made from the intestines (6), and heparin, for treating blood clots, is taken from the small intestine (7). The gall bladder (8) yields dyes and insulin is taken from the pancreas (9).

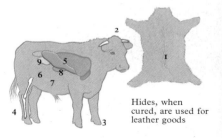

Hides, when cured, are used for leather goods

Shows Agricultural shows are an important feature of the livestock breeder's calendar. Prize-winning enhances the value of the breeder's stock and exhibitors spend much time and effort preparing and grooming their animals for showing. Judging is done in classes by experts such as farmers and breeders, with each breed divided into different classes according to age and type. Animals are normally judged by their conformation, though points are sometimes awarded for their level of productivity. Breed and Reserve Champions are selected from the winners of the classes, and a Supreme Champion is selected from all the first prize winners at the show.

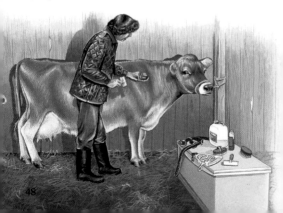

Sales Pedigree cattle are normally sold separately from non-pedigree or commercial stock. They are usually sold at auctions, often organized by a Breed Society who will prepare an official catalogue with full details of the pedigree and performance of the animals. Auctions are also an important aspect of buying and selling commercial cattle, particularly beef calves and store cattle. Finished beef cattle are either sold privately or by auction to butchers and abattoirs. Surplus dairy cows are usually sold for slaughter but are sometimes auctioned to other farmers.

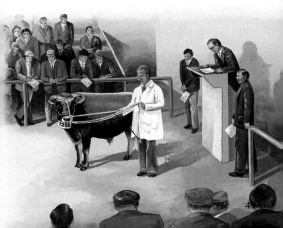

Deer

The renewed interest in deer farming in Britain has been encouraged by the demand for venison, particularly from Europe, and by the knowledge that, with improved feeding, deer can grow rapidly in enclosed parkland. Three kinds of deer are commonly farmed, the Red Deer, the Fallow Deer, and the Roe Deer.

The food consumed by deer in the wild is of low digestibility – only about 25–30 per cent of the tough plant fibres eaten are useful – and because of its shy nature the deer has to eat and swallow fast, without chewing. It is the ability to regurgitate and rechew woody plant fibres in a safe spot at leisure, using its powerful cheek teeth to shred and grind its food, that allows the wild deer to survive and grow.

Deer have a short mating season in late summer and autumn. Oestrus occurs every 18 days, although the Roe Deer may show oestrus only once in a season. Pregnancy normally lasts for 230–235 days, the calves being born in late May and early June. Fallow and Red Deer average one calf of about 6.4 kg, but the Roe Deer frequently has twins, each of 2 kg. The hinds are ready to breed at 15–18 months of age, and in enclosed parkland a stag can cover 10–20 a season.

Red Deer in a fenced enclosu

Production Deer must be kept in specially fenced enclosures, with fencing of at least 2 m high and well-secured at the base, otherwise they will leap over or squeeze through it. In these conditions, Red Deer will grow from birth to 90 kg in 14–16 months, with the stags growing faster. Young Red Deer receive milk with a high protein and mineral content and a large proportion of solids (34.1 per cent), compared with only 12.5 per cent in the dairy cow. They can maintain a growth rate of 200 g per day from birth to sale, which compares very favourably with young lambs growing at that rate from birth to five months. Deer need to be kept at about eight per hectare to produce the same meat per hectare as a flock of sheep, but so far stocking rates of only 2–5 per hectare have been maintained. This could be improved by keeping them on richer pasture. Red Deer produce the most expensive venison, although it may be less gamey than that of wild animals.

Deer species The Red Deer is the largest British deer and the easiest to tame, and so is the most suitable for farming. The stag weighs 140–160 kg and stands 1.2 m at the shoulder; the hind weighs rather less. The dominant stag must be approached with care at all times, particularly in the breeding

Red Deer stag belling in the autumn rut

season. Red Deer are found in the wild on the open moorlands of Scotland, the Lake District, Somerset and Devon.

Antlers, which only occur in the stag of deer species, are shed in April and regrow to full size by the autumn. Those from a mature Red Deer may weigh 9 kg. Other by-products of slaughtered animals include skins, tails, sinews and pizzles.

Fallow Deer

Mature antlers
of the Fallow
Deer:

1 Palm points
2 Guard tines
3 Trez tines
4 Brow tines

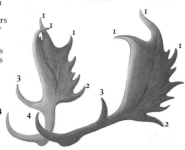

The Fallow Deer is intermediate in size between the Red and Roe Deer. The stag reaches 90 cm at the shoulder and weighs 90 kg, the weight of a large ram. There is considerable variation in colour within the species, from light brown to black. Fallow Deer are found in woods and forests throughout England and Wales, but are less suitable for farming than the Red Deer because they are harder to domesticate.

The Roe Deer is the most difficult to farm, despite its ability to produce twins, because of its shy and solitary nature. The stag stands 75 cm at the shoulder and weighs 30 kg; the doe is rather smaller. Like the Fallow Deer, it is found in well-wooded areas of England.

Roe Deer

Sheep

Sheep have been used by man for thousands of years for their wool and meat – the white-faced breeds of sheep seen in Britain today are believed to be descendants of those brought over by the Romans. There are now 30 million sheep in Britain (counting lambs as well as ewes and rams), or one for every two people, but their place in British agriculture is not as important as it used to be. The emphasis has changed from the sale of wool to the sale of lambs, for nowadays a farmer makes ten times as much money from a ewe's lambs as from her fleece.

A hundred years ago, on the Downs of southern England, sheep were driven out from the villages every day to pasture (pictured opposite), often walking three miles in the morning and back in the evening.

Lincoln Longwool
The improved 19th-century sheep resembled this breed

Upland sheep Methods of keeping sheep have changed, but the upland environment has not. Mountain sheep are hardy, defying snow, gales and biting cold; they stay alive where cattle would die, cropping the fibrous heather and coarse grasses. To encourage the use of mountainous areas a subsidy is paid by the government for every ewe kept. The most common mountain breeds are the Scottish Blackface and the Welsh Mountain. At lower altitudes sheep graze in company with cattle, for upland areas are less bleak than the mountains: grass grows for a longer period of the year, the ewe has more to eat and can rear twin lambs without too great a strain.

Lowland sheep The most common ewes to be seen on the lowlands are crossbreeds. These are bred in hill and upland districts by mating a prolific ram, such as the Border Leicester or Teeswater, with a mountain ewe. The crossbred ewe, if fed well prior to mating, will produce mostly twins or triplets. More grass is grown on the lowlands, and so from each hectare the farmer may keep some 15 ewes and sell up to 30 lambs – in contrast to the mountain regions where often only one ewe and lamb per hectare can be kept and the lamb will still have to be fattened on the lowlands. Quite often lowland sheep are kept on specially sown pastures.

Buying sheep Sheep are normally bought at market in the autumn, before mating takes place. It is an apt time, therefore, to begin the shepherd's year. Each breed or crossbreed has its own traditional markets or fairs: Southdown sheep from Findon, Sussex; Kerries and Cluns from Craven Arms, Shropshire; Welsh Mountains from Builth Wells and Caernarfon; Scottish Halfbreds, Greyfaces, Mules and Mashams from Kendal and Lancaster; and north of the Border, Scottish Blackfaces and Cheviots from Lanark and Newton Stewart. Ewes of any age can be bought in lots of 20 or more, from ewe lambs to broken-mouthed ewes sound in udder but not in tooth and of uncertain age. The name for ewes of different ages changes with the region: eighteen-month old sheep can be called gimmers, theaves, shearlings or, more generally, two-tooths. Younger sheep are easily aged by their front incisor teeth, having two large central incisors at 16 months, four at 22 months, six at 28 months and eight (full mouth) at three years. From then on it is not so easy.

Rams must also be bought, usually at auction, and top pedigree rams can command huge prices. The most common ram breed for prime lamb, the Suffolk, is shown opposite. Rams tend to have a shorter breeding life than ewes.

Flushing and mating After the farmer has brought his flock up to strength at the sales, the ewes are 'flushed' prior to mating. This means grazing the flock on good grass to increase their weight. Broadly speaking, the fatter a ewe is at mating time the more lambs she will have, though, for instance, a fat Masham will have more lambs than a fat Welsh Mountain, because some breeds are more prolific than others. The number of lambs born at one lambing may reach as high as 270 per 100 ewes.

Sheep have a limited breeding season when they will stand for the ram, its time coinciding with the shortening of daylength in the autumn. Thus ewes in the northern hemisphere tend to breed between September and February, while those in the southern hemisphere do so between March and August, although some breeds such as the Dorset Horn and the Merino can breed almost all the year round. A ewe will show oestrus for about 24 hours and if not mated will show oestrus again 16 days later. Rams are normally expected to mate about 40 ewes a season in Britain.

Mating behaviour is shown opposite. The ram goes from ewe to ewe and sniffs each one's tail region, curling his lips in a characteristic manner. When he finds a ewe in oestrus he nudges her with his head or front foot; if she stays still he mounts her.

To help the shepherd monitor his flock at mating time, a ram can either be painted on its chest or harnessed with crayons of different colours (yellow, blue, red or green – shown opposite) that rub off on the ewe at mating. The colour is changed every fortnight and the shepherd then knows which ewes will lamb first and the flock can be divided on a colour basis and fed accordingly

63

Winter feeding As autumn approaches the grass stops growing and its quality falls with the greater amount of old and dead leaves that are present. From now on, it is necessary to supplement the ewes' diet with hay or concentrated feed blocks, particularly from January to March when deep snow or thick mud may prevent ewes wintering outside from getting to their normal feeding places. A ewe mated in October or November will lamb in March or April, and towards the end of pregnancy needs more food to nourish the growing foetus. It is essential that she eats daily, otherwise the ewe and her unborn lamb will die from pregnancy toxaemia (twin lamb disease).

Housing If possible, ewes should be housed during the worst of the winter in sheds with good ventilation but with no draughts at sheep height. They can be fed hay or silage and a little concentrate (mostly barley), increasing in amount as lambing gets closer. They are usually bedded on straw, peat or slats in groups of 20–25 with adequate feeding and lying space, otherwise the more timid ewes will be unable to get sufficient food. By housing his sheep the shepherd can feed and watch them more easily and it prevents the pastures from becoming too muddy, so allowing earlier growth of grass in the spring.

Lambing Pregnancy lasts for about 147 days in sheep, and the lambs are usually born in March or April. Just before lambing the ewe will go off on her own, either in the field or indoors in a pen, and paw at the ground. Her udder becomes larger and milk can be drawn from the teats.

During lambing the ewe should be watched closely but not interfered with unless the lamb is very slow to emerge, usually because it is the wrong way round or has too large a head – the head should come first, together with the two front legs. Quite often in difficult cases the shepherd can rearrange the lamb inside so that it emerges normally. After the lamb is born the ewe will lick the membranes and birth fluids from it, enabling it to breathe and imprinting its smell in her mind. Strong lambs get up within minutes of birth and nose their way towards the udder.

After lambing, the ewe and her lamb or lambs should be put into a pen on their own. This prevents other ewes from trying to steal or adopt the lambs and gives the family time to establish its bonds. After 24 hours, if the lambs are feeding properly and have had the first milk which is rich in protective antibodies and is called colostrum, the family can be moved to a larger pen or out to pasture. Great care has to be taken at this stage to avoid panicking the ewes and getting the lambs separated from mother.

Fostering Inevitably some ewes refuse their lambs, either because they are inexperienced mothers or else because ewe and lamb have become separated shortly after birth. A rejected lamb should quickly be fostered onto another ewe whose own has died, or which has had only a single lamb. To encourage the ewe to adopt another lamb, the shepherd will put her dead lamb's skin on it or rub her birth fluids into its coat. Fostering crates are available for the most difficult cases. Here the ewe is fastened by the head to prevent her from butting the lamb away. As it suckles and lies by her for a day or two she is then more likely to accept it.

Fostering crate

**astrator
d rings**

Docking Within two or three days of birth, and provided the lambs are strong enough, the shepherd will put elastrator rings on their tails and castrate the males with similar rings. Lambs' tails are docked for reasons of hygiene and subsequent health, for lambs with long tails are more likely to get mucky at the rear when they start to 'scour' when fed with young grass. This encourages flies to lay their eggs in the soiled areas and very quickly the lamb is 'struck' and literally eaten alive by the emerging maggots. About three or four weeks after ringing, the tails and purses drop off.

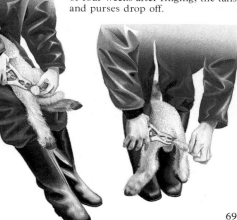

Spring grazing After lambing, the ewes and their lambs are put out to graze the lush spring pastures. Sometimes the ewes will 'scour' after changing over from dry hay indoors to leafy young grass and they are also in danger of getting grass tetany if they do not receive enough magnesium in their diet. The ewes produce most milk about two or three weeks after lambing and their appetite soon reaches double that before lambing. In Britain most ewes will lactate for 12–16 weeks.

Some farmers graze their sheep in one field throughout the year, some use several fields rotationally and some graze their ewes in one field and allow the lambs to creep through specially constructed hurdles into the next field in the rotation. By three weeks of age the lambs have started to nibble grass, thus stimulating development of the rumen.

Wool production The woollen cloth of Britain has been famous for its fineness for hundreds of years, and during the Middle Ages wool was the country's greatest asset. Production was concentrated in the West Country and in the Pennine districts of Yorkshire and Lancashire, with their plentiful sheep pastures, soft water for washing, scouring and dyeing, and water power to drive the mills. During the Industrial Revolution, however, wool manufacture became centred upon Bradford, for this was the area least hostile to mechanization.

Shearing (pictured opposite) is a highly skilled craft and an expert can shear a sheep in 60 seconds. The farmer's aim is to keep the fleece free from dirt, straw and paint marks, and the shearer's aim is to keep the fleece whole.

Rolling a fleece ready for tying

Merino
Renowned for its high yield of wool, with skinfolds to increase its surface area

A skilful shearer holds the sheep firmly with his legs while cutting away the fleece from the head towards the tail

73

Unimproved wild sheep have a double coat (see diagram), with a layer of coarse kemp (1) protecting an underlayer of fine wool (2). They shed their coat completely each spring, unlike improved breeds, and also have undesirable fleece pigments which are resistant to dyes. Sheep have, therefore, been bred to be either white or black and for only one type of fleece. Although British breeds do not clip the same weight of fine wool as Merinos, which may yield 16 kg of fleece, they do provide a useful, wide range. The nearest British breed in quality is the Southdown, but it only clips 2.5 kg of fleece.

Raw wool (shown opposite) is divided into fleeces of different fineness counts, known as the Bradford count, which is based on the number of yarn lengths, each of 560 yards, that can be spun from one pound of wool prepared for spinning. Merino wool is the finest, with a count of over 60 yarn lengths, and is made into the best quality light worsted and woollen garments. Crossbred wools (44–58 yarn lengths), from most British sheep, are used for good tweeds, lower quality worsteds and knitting wool, and finally the coarsest springy wools (below 44 yarn lengths) are made into carpets and mattresses. This would be wool from Scottish Blackfaces, Swaledales and Devon Longwools.

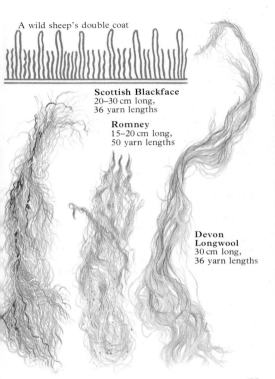

A wild sheep's double coat

Scottish Blackface
20–30 cm long,
36 yarn lengths

Romney
15–20 cm long,
50 yarn lengths

Devon Longwool
30 cm long,
36 yarn lengths

Health Sheep get their fair share of ailments, both internal and external, many of which can be avoided by nutrition, hygiene, preventive medicine and the keen eye of the shepherd. Vaccinations and injections can prevent many of the more spectacular diseases, but there are still some, such as scrapie, where the animal rubs itself continually against posts and gates, getting thinner and thinner, which have defied cure. Some deaths are accidental, such as ewes getting 'cast' – stuck on their backs with their legs in the air and unable to roll over onto their side and thence onto their legs. They die of suffocation when the contents of the rumen spill into the windpipe.

A number of farmers breed their own replacement stock because of the risk of buying in sheep with diseases such as contagious abortion, which may strike catastrophically during pregnancy and kill great numbers of lambs. Lambing is the most common time of loss, and each year millions of pounds are lost to the industry through lamb mortality. Mastitis is as frequent in sheep as in dairy cows but receives much less attention – lambs get too little milk, become 'bad-doers' and are then much more likely to succumb to diseases and parasites. A lumpy udder from mastitis is a common reason for culling ewes.

Undernutrition in pregnancy and lack of certain elements such as magnesium and calcium during lactation, and of trace elements such as copper and selenium, can also cause deaths.

Land that has had sheep grazing on it for year after year becomes 'sheep-sick' and the stock never do well on it again until it has been rested for a year or more.

Before

After

A sheep's hooves need to be frequently trimmed back or they will become overgrown and misshapen, and infected with foot-rot. After trimming, the sheep is put through a foot-bath containing formalin or copper sulphate and left to dry on clean concrete

Sheep are beset by many different kinds of internal parasite which damage their liver, lungs, stomach, intestine and brain. The life history of one of them, liver fluke, is shown opposite. This has a secondary host, a small snail which lives in water. If wet marshy pastures are drained then the snail is discouraged and liver fluke is no longer a problem.

Parasites can cause scouring, emaciation and death, particularly among lambs. In the spring, the ewes deposit vast quantities of parasite eggs onto the pasture and when these hatch out, the time of year depending on the kind of parasite, the infective larvae are eaten by the lambs along with the grass. The best solution is to keep the lambs on clean pasture and to dose the ewes with an anthelmintic or worm drench (pictured opposite) before they are put out to graze in the spring. Provided that the doses of parasite larvae that the lamb is challenged by are small, and provided that it is getting plenty of milk and grass, it will build up a resistance to the parasites.

The most common time for 'illthrift' in lambs is late July and August, when there are most infective larvae on the grass. The lambs should therefore be weaned on to clean pasture in late June. If this is impossible then the only alternative is repeated dosing.

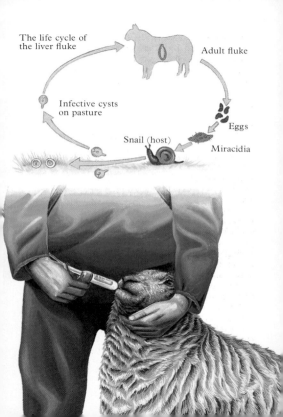

The life cycle of the liver fluke

Adult fluke

Infective cysts on pasture

Eggs

Snail (host)

Miracidia

Dipping Sheep are dipped to kill the commoner external parasites such as keds, lice and ticks, to prevent fly strike, and in some areas to prevent the spread of sheep scab. Keds and lice are bloodsuckers and greatly irritate the sheep while at the same time causing a loss of vitality and soiling the wool. Fly strike occurs most frequently during warm thundery weather: the blow flies lay eggs in the soiled parts of the fleece and then as the maggots hatch they burrow under the skin of the animal. A 'struck' sheep appears very uneasy, continually turning round and waggling its tail. Once the maggots hatch a ewe or lamb will die in two or three days if not treated. Sheep scab is a form of mange. The progress of the disease is very rapid and it is one of the worst of sheep afflictions. It is notifiable in Britain. Another scourge of upland and mountain sheep is the head fly.

Sheep dips are effective if used at the right concentration. The picture opposite shows sheep being dipped, and the secret is to keep the flow of sheep into the dip moving while ensuring that each animal is thoroughly immersed for at least 30 seconds. After dipping, the ewes are left to stand and drain for 15 minutes, because the chemicals in the dip would contaminate the pasture.

Pasturing Ewes make good scavengers, grazing over cereal stubbles, eating out weeds and returning nutrients to the soil. So useful is this that in countries where sheep flocks are still nomadic the shepherd is paid to graze his flock over the stubbles, providing that he leaves them there at night so that their droppings can further enrich the soil. 'Where the sheep's foot treads, there the land turns to gold.'

During the long summer days the lambs thrive and fatten on pastures of rich new grass, playing in groups, leaping from tree stumps, racing up and down banks. They return to suckle at ever-lengthening intervals and as the lambs grow the poor ewe gets lifted off her hind feet whenever her twins bunt her udder for milk. June and July pass, the grass slows in its growth and where there was plenty there is now competition for the coarser, thicker, shorter blades. It is several months yet before the ewe has to be in good condition for mating; her milk is drying up and the farmer is looking forward to money from selling prime lambs. He therefore weans the lambs and gives them the best and cleanest of the remaining grass ('clean' from infective parasite larvae). If the lambs get checked in their growth they grow leggy and lose condition, and will need several more months before they are fit for the butcher.

Fattening lambs The majority of lambs are sold as prime fat lamb in August and September. The singles go first because they have had the most milk, then the twins and finally the triplets. Various grazing systems are used to fatten them, ranging from the simplest – set stocking, where the ewes and lambs graze the same area throughout the season – to more elaborate systems of rotational grazing with six paddocks and forward creep for the lambs. Here the lambs squeeze through specially constructed gates (creeps) into the best grass and the ewes remain behind. The creep gate should be opened when the lambs are 2–3 weeks old and still at the inquisitive stage. This method ensures even defoliation, no wasted coarse patches, and at the same time allows the lambs to grow quickly on the choicest grass. It also removes the need to wean and has been shown to promote the fastest lamb growth.

Those lambs that are not fattened off grass by September and October can be grazed on specially sown crops such as rape, stubble turnips or swedes. Such lambs are known as store lambs, or 'stores'. They are usually folded across the crop with a ration for a week, fenced off with electric fencing (strip-grazing, shown opposite); otherwise the whole field can get muddy in bad weather and the leaves will be trampled.

85

Lamb production Because of the large number of breeds and crosses in Britain, lambs come in all shapes and sizes. Indeed, it is a criticism of British lamb that it is not a uniform product compared with, for example, New Zealand lamb which is practically always from one breed, the Romney.

In an attempt to produce lean carcasses of uniform size, a grading scheme has been introduced which guarantees not only the shape but also the amount of fat cover. The price structure, too, is now changing in favour of the lighter, leaner carcass. In the past, carcasses tended to be larger and fatter, because the more a lamb weighed the more money a farmer received for it, even though he was paid less per kg for carcasses with excess fat. In the nineteenth century, sheep meat was usually much fattier and older, but that was mutton and not lamb.

Lambs are chosen for sale by feel, in particular of the dock and loin area. Batches of lambs are then sent either for sale 'on the hoof' or else to abattoirs 'on the hook'. A premium is paid if a high proportion of those sent meet the required specification. Any lamb that is not finished well enough or has poor conformation is classed as a reject and does not receive a subsidy. Meat inspectors scrutinize all carcasses and those with excess bruising, lungworm or liver fluke are condemned.

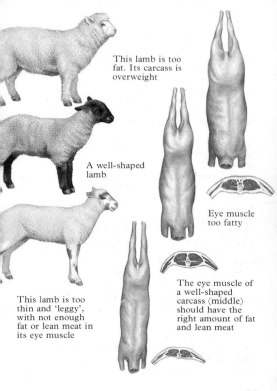

This lamb is too fat. Its carcass is overweight

A well-shaped lamb

This lamb is too thin and 'leggy', with not enough fat or lean meat in its eye muscle

Eye muscle too fatty

The eye muscle of a well-shaped carcass (middle) should have the right amount of fat and lean meat

87

Meat The carcass weight of a lamb is just under half the live weight, so that lambs of 36–38 kg live weight will make carcasses of 16–18 kg. The butcher pays for the carcass on grade and weight, which means that any excess fat he has to trim off before dividing it into cuts is money wasted, for the fat has no resale value. Lighter carcasses are now more in demand, and are more expensive, both because they contain less fat and because smaller joints have become more popular. The division of the lamb carcass into the various cuts is shown opposite.

A lamb carcass of good quality should have short bones and well-filled legs which are U-shaped rather than V-shaped, with a covering of fat carried down to the hocks so that the meat does not dry up in cooking. The loin, which is the most valuable cut, should be wide, well-filled and have a deep eye muscle, and the ribs should be short and, like the neck, light, for these are both low-priced cuts. If farmers know the carcass grade and weight required by the meat trade, they can breed and fatten their stock accordingly.

When looking at lamb in the shop the younger the animal the paler will be its flesh. The difference between imported and home-produced lamb is in the colour of the fat, with the former being firm and white and the latter creamy coloured.

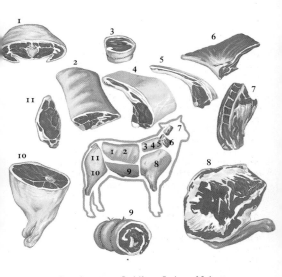

Lamb cuts **1** Saddle, **2** Loin, **3** Noisettes,
4 Best end of neck, **5** Cutlets, **6** Middle neck,
7 Scrag, **8** Shoulder, **9** Rolled breast, **10** Leg,
11 Chump chops

Culling Each year the ewes are sorted out and those thought no longer able adequately to rear a lamb are sold for slaughter. This normally takes place in the autumn before the flock is put to the ram; the shepherd's year has then gone full cycle.

The main reason for culling is lack of condition, often caused by hidden ailments. Other reasons are persistently bad feet, lumpy udders from mastitis, damage to the teats, and a tendency to prolapse at lambing – this is the pushing out of the vagina and uterus when heavily pregnant. Sheep that have been twice barren may also be culled. Some degree of flock recording, particularly at lambing time, is a great aid to the removal of unproductive or troublesome ewes.

Another common reason for culling is lack of front incisor teeth. These ewes are called 'broken-mouthed', although opinion is divided as to how important a full set of incisors really is for biting and chewing grass and hay. At present, sheep are often culled when they start to lose condition at six years of age. There is no sign, however, that the number of lambs they have or their milk yield declines until they are about nine years old.

Sheep only have front teeth in the lower jaw; there is a hard pad on the upper jaw. They start life with milk teeth and at 12–16 months of age the permanent incisors start to come through, the two centre ones first, then two more about every eight months – a full mouth at three years of age. After this, depending on nutrition, the permanent incisors get worn down, grow too tall or fall out. See the illustrations below.

Besides the eight incisors a sheep has 24 cheek teeth, 12 on the lower jaw (six on each side) and 12 on the upper jaw. These are quite hard to see because a sheep does not open its mouth very widely without outside pressure. There is a gap of about 5 cm between the last incisor and the first of the premolars in the lower cheek.

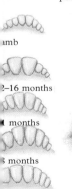

amb

2–16 months

4 months

3 months

years

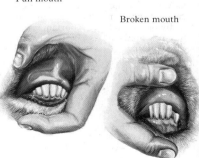

Full mouth

Broken mouth

Breeds Identifying breeds of sheep is not easy in Britain. There are over 40 pure breeds and, more confusingly, a great many crossbreeds, not just first crosses but second crosses made partly for convenience – for instance, the Suffolk on the Scottish Half-bred. In addition, the amateur breeder rightly takes a great interest in improving his stock and will go to great lengths to fix his own ideal 'breed'. Foreign importations, such as the Finnish Landrace and the East Friesland, are also made from time to time to improve our native stock and are used for crossing. The resulting crossbred flock of sheep may well look and be very different from their forebears.

The descriptions and illustrations which follow will aid recognition of some of the pure breeds and more usual crossbreeds found in Britain. Numerically there are more crossbred than purebred ewes. The important points to concentrate on are the colour and shape of the face; the position of the ears and the presence of horns; the type and appearance of the fleece; and the size, though this can be rather deceptive if the fleece is long.

The two breeds pictured opposite are the two most common mountain breeds, the Scottish Blackface and the Welsh Mountain.

Scottish Blackface
Both ewe (left) and
ram are horned

Welsh Mountain
Ewe (left) and ram

93

Mountain breeds The Scottish Blackface (p. 93) is the most numerous mountain breed in Britain with 1.6 million ewes, most of which are found on the mainland of Scotland. It is renowned for its hardiness, and, although producing one lamb in the mountains, the ewe (53 kg) can average 1.7 in the lowlands. The breed type differs according to the district.

The Welsh Mountain (p. 93) is the second most numerous breed with 1.3 million ewes, and is found mainly in the mountains of Wales. It is a small, active sheep weighing 35 kg and rarely gives birth to more than a single lamb. Like the Scottish Blackface, the type varies with the region.

The relatively large Cheviot (73 kg) was introduced into Scotland from northern England over 600 years ago. It is more of a hill than a mountain sheep and there are two types, the South country and the North country, which has finer, silkier wool.

The distinctive Herdwick has been in the Lake District for almost a thousand years. Its thick, cold-resistant fleece was originally made into 'hodden grey' cloth.

The Swaledale (50 kg) can be distinguished from the Scottish Blackface by its lighter-coloured mouth. It is a hardy sheep, popular in northern England, with coarse wool used for making carpets and tweeds.

Cheviot
White-faced,
hornless and
fairly large

Herdwick
Hornless,
and grey in
colour

Swaledale
A small and
active breed

95

Crossing breeds of ram The three breeds pictured opposite, together with the Wensleydale, when mated with mountain or hill ewes, produce a great range of popular halfbred ewes for the lowlands and uplands. The rams pass on size and prolificacy to their offspring, while the mountain ewes pass on hardiness and milking ability.

The Border Leicester is primarily from the Scottish Lowlands, where it is descended from the Cheviot and the Leicester. It is most usually crossed with the Cheviot to produce the Scottish Halfbred, with the Scottish Blackface to produce the traditional Greyface and with the Welsh Mountain to produce the Welsh Halfbred.

The Teeswater from north-eastern England produces the popular Masham when crossed with the Swaledale or Dalesbred ewe. When crossed with the Welsh Mountain it produces the Welsh Masham. The Wensleydale resembles the Teeswater, but has a blue-grey face and a slightly longer ringletted fleece. It is also used on the Swaledale.

The strange-looking Blue-faced or Hexham Leicester produces the Greyface when crossed with the Swaledale. When crossed with other hill or mountain ewes it produces a Mule, and when crossed with a Welsh Mountain it produces a Welsh Mule.

Border Leicester
An aquiline face and large pricked ears

Teeswater
Brown or black markings on a white face and a long curly fleece

Blue-faced or **Hexham Leicester**
Curved, boney nose and a very long back

97

Crossbred ewes When mountain ewes have produced three or four lambs they are sold to an easier area and mated with a crossing breed of ram. The resultant crossbreeds are then sold to the lowlands, usually at 18 months of age, and mated with 'terminal sires', mostly Down rams, to produce the bulk of the prime lambs that are eaten in Britain. The 6.5 million crossbred ewes on the lowlands are the heart of the British sheep industry. The four most common are shown opposite.

The Scottish Halfbred is a big (77 kg) white-faced ewe and averages 1.76 live lambs born per ewe lambing. The lambs are large, often weighing 5.5 kg each.

The Welsh Halfbred is smaller (58 kg), with a less pronounced curve to its nose, and does not produce so many lambs (lambing average 1.56).

The distinctive, black-and-white faced Masham weighs about 70 kg and is very prolific (lambing average 1.79). It produces a large number of triplets.

The Mule is a popular halfbreed, daintier than the Masham and with a shorter fleece. It is a similar size (73 kg) and equally prolific.

A common second cross ewe is from a Suffolk ram crossed with the Scottish Halfbred.

Scottish Halfbred
A hardy and prolific breed

Welsh Halfbred

Masham
Distinctive head, body and fleece

Mule (right)
Greyer face than the Masham

99

Prime lamb sires The purpose of the prime lamb sire is to produce offspring that will grow fast and mature into lean lambs of good conformation – that is with more muscle tissue and less fat.

The Dorset Down has the characteristic broad head, wide-set ears and dark face with wooly sideburns of most Down breeds. A mature ram will weigh 100 kg and produce lambs of good conformation at 36–38 kg live weight at 12–16 weeks of age. Other popular Down breeds are the Oxford, the Hampshire, the Shropshire, the Southdown and the Ryeland.

The Suffolk ram, which is also a Down breed, is the most widely used prime lamb sire – 34 per cent of all ewes are mated to Suffolks. It has a very black face and a short fleece, and its cross lambs command good prices as store lambs because they mature into lean, heavy lambs of about 45 kg during the winter.

The Texel is from Holland and many hope it will have the kind of impact on lamb production that the Charolais has had on beef cattle. Its offspring do not grow any faster than those of the Suffolk or the Dorset Down, but their carcasses have a higher proportion of lean meat. Other European importations that have been tried as prime lamb sires include the Île de France (p. 104) and the Oldenburg.

Dorset Down
Typically stocky Down breed

Suffolk
Very popular prime lamb sire

Texel
Promising prime lamb sire from Holland

101

Purebred lowland ewes These serve a similar purpose to crossbred ewes, but have a strongly regional distribution. Different breed types have been bred in and for a particular area.

The Clun Forest is found mainly in Shropshire, Herefordshire and other Welsh Border counties. It has a distinctive black face, a neat, hornless head, short pricked ears, a concave profile and a tufted forelock. The ewe weighs 62 kg and is quite prolific (lambing average 1.73). The Kerry Hill is found in neighbouring areas to the Clun.

The stocky and docile Dorset Horn, which can be horned or polled, has a broad white face and short wool. It weighs 72 kg, with a lambing average of 1.53. Its long breeding season from July to March makes it useful for producing early lambs for the Easter market and for lambing more than once a year.

The white-faced Romney Marsh or Kent ewe thrives on the short swards of the Romney Marsh at high stocking rates. The ewe weighs 69 kg, produces a lambing average of 1.36 and clips 4.5 kg of wool.

The large (89 kg) Devon Longwool, or Devon and Cornwall Longwool, has a long fleece and a docile temperament. It produces a lambing average of 1.54. Related breeds in the area are the Dartmoor and the Devon Closewool.

Clun Forest
A prolific
purebred ewe

Dorset Horn
Useful for
early lambs

**Romney
Marsh**
Thrives in
exposed
conditions

Devon Longwool
(left) A docile,
popular West
Country breed

103

European Breeds

Galway
Comes from
western
Ireland

Île de France
(right) A sire of
prime lamb

East Friesland
Milking ewe from
West Germany

Finnish Landrace
A prolific breed
from Finland

Ancient Breeds

Jacob
This breed is
mentioned in
the Bible

Shetland (above)
May be white,
brown or black

Soay
A small breed
with long legs

Manx Loghtan
Sometimes has
six horns

Sheep milking During peak lactation, about 3–4 weeks after lambing, a ewe can produce 3–4 litres of milk per day. This is less than a dairy cow or goat, but is still a useful yield in relation to the sheep's size and appetite. The picture opposite shows a Lacaune ewe being milked by hand. Sheep milking is much commoner in France, Spain and Italy than in Britain. France has a number of sheep's milk cheeses, the most famous being Roquefort, a blue cheese that is matured in caves. Some typical cheeses are shown here.

British breeds of sheep can yield up to 173 litres of milk in 12 weeks, although a specialist milk-producing ewe, such as the East Friesland (p. 104), can produce almost ten times this amount, but in a longer lactation. A ewe suckling twins will produce more milk than one with only a single lamb, because more of her milk is being withdrawn daily and suckling is more frequent. The same is true of a ewe rearing triplets, but the increase in milk yield is less between twins and triplets than between singles and twins.

Sheep's milk is higher in solids than cow's milk, and so if an orphaned lamb is being hand-reared it needs a milk powder which is more concentrated than ordinary cow's milk. Orphan lambs need feeding little and often.

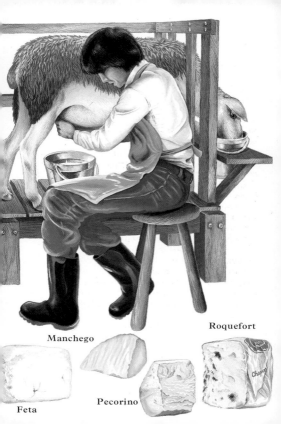

Manchego

Roquefort

Feta

Pecorino

Sheepdogs

Without the working sheepdog there would be no domesticated sheep on the hills and mountains of Britain, for one shepherd with his dog or dogs can gather more sheep than a whole host of men on their own. Mountain areas can be rugged and treacherously boggy, but whether in the mountains or the lowlands sheep have to be looked at regularly to see that they are healthy, and gathered in for shearing, dipping and sale. A strong working dog is therefore a necessity. It must be fit and have instinctive intelligence.

Shepherds originally kept dogs to guard against predators and thieves: large breeds such as the Old English Sheepdog, the Italian Maremma and the Pyrenean Mountain Dog were widely used. But today there are no large predators to be kept at bay in Britain, and so it is the fitness and intelligence of the working dog that are at a premium, and not its strength and ferocity.

The familiar black-and-white sheepdog, the Border Collie, stands about 56 cm at the shoulder and weighs 20 kg. It is expected to cover 100 miles (161 km) in a ten-hour day. The breed was known in the seventeenth century in Scotland and in the eighteenth in northern England, and it is now the most widely kept sheepdog in the world.

Sheepdogs are frequently used to gather up a group of sheep so that the shepherd can see how the sheep move and whether any are badly lame or fly-struck.

110

Training It is a much-debated point as to how much the shepherd trains the sheepdog and how much the dog influences the man, but without question a very close relationship quickly grows up between the two, such that their understanding becomes instinctive. The sheepdog has to dominate the sheep so that it can move them exactly where it wants to and not too fast, particularly with heavily pregnant ewes and those with young lambs. It should be able to do this by eye and by presence and with a minimum of biting, but the herding instinct is closely allied to the hunting instinct and part of the fascination of watching a sheepdog work is the way its ferocity is used inwardly to dominate but not outwardly to kill. The sheep has to feel that it may at any moment be gripped by the dog but the dog has always to pretend, most importantly to itself. Some dogs are bold and impetuous, others are much more gentle.

The instinct to move other animals shows as early as six weeks in sheepdog puppies, but they are not normally trained and worked until eight months of age.

There are five basic commands: to stop (*lie down*), to move on (*come in*), to go right (*away to me*), to go left (*come by*) and to finish (*that'll do*). The actual words vary between shepherds and whistling is often used instead.

Trials Sheepdog trials play an important part in emphasizing successful training, and the way one or two dogs move a small group of sheep between posts, help to single out one sheep, and pen and hold the rest is fascinating to watch. These dogs and their offspring can command huge prices, but trial dogs are not necessarily well-suited to working on a commercial sheep farm where there are large numbers of sheep to be moved.

Legends of the sheepdog's intelligence and loyalty abound, but it is the sheepdog's importance to the sheep industry, particularly in the remoter parts of the country, that makes it so popular. By working it is fulfilling its natural role.

Goats

Goats account for 15 per cent of the world's domestic grazing animals, although in Britain only 6000 were registered at the latest census. This is small compared with the 80 000 in Switzerland, the 960 000 in Italy and the 1 048 000 in France. Goats have been selectively bred for four principal characteristics: milk yield, milk protein yield (which affects cheese production), ease of milking and early maturity.

There are four main breeds in Britain: the British Saanen, the British Toggenburg, the British Alpine and the Anglo–Nubian. The 'British' goat is a crossbreed and as such has no fixed appearance. Because the line of the nose, the set of the ears and the colour of the coat are irrelevant in a crossbreed, it has been possible to concentrate on milk yield and conformation, which is why the majority of champion goats have been 'British' goats and why there are more crossbred goats than of any pure breed.

The great improvement in the milk yield of goats kept in Britain in the last 100 years has derived from Switzerland, for it was in the Alpine valleys that the world-famous breeds of goat, the Saanen and the Toggenburg, were originally selected and fixed. The first Toggenburgs were imported into Britain in 1884 and the Saanens in 1903.

British Saanen
The most popular breed and a good milker

British Toggenburg
Famed for its high yield of milk

Goat breeds The Saanen originated in the Saane and Simmental valleys of the Berne region of Switzerland. It is an outstanding milker (2870 litres per year) and is white, pale cream or pale fawn in colour, with black spots on the nose, ears and udder. The British version of the Saanen (female 68 kg, male 91 kg) is bigger than the Swiss with stronger hindquarters, but retains the placid nature that makes it eminently suitable for an indoor life.

The Toggenburg came from Obertoggenburg and Werdenburg in north-east Switzerland. The colour is fawn or light brown, with white markings. As with the Saanen, the British Toggenburg (female 63 kg, male 80 kg) is larger than the Swiss and darker in colour, and is almost as good a milker as the British Saanen.

The British Alpine can be described as a black-and-white Toggenburg, from which it differs mainly in colour, and which originated in the Swiss and Austrian Alps.

The Anglo–Nubian has the distinctive lop ears and aquiline nose of the Nubian and is a bigger animal (female 73 kg, male 95 kg) than the breeds of Swiss origin. It can be various colours but white and roan predominate. It is the best for meat but is not so good for milk, although the butterfat content is high. It is noisier and more temperamental than the Swiss breeds.

British Alpine
Noted for its fine
glossy coat

Anglo-Nubian
A large breed with
distinctive ears and
face

Breeding and feeding Goats, like sheep, are seasonal breeders when kept in temperate latitudes, breeding from September to January. The oestrous cycle is 21 days long and heat can last 2–3 days. Pregnancy lasts on average 150 days. The number of kids born varies with breed but twins are the most common. The British Saanen averages 1.9 kids per litter, while figures for the Anglo–Nubian range from 1.5 to 2.3. There is some infertility with bucks, particularly in hot climates.

About 6–7 per cent of goats are hermaphrodites. These hermaphrodites are genetically female at conception but become intersexes during the development of the embryo. Hermaphroditism is linked with the gene for the polled condition, so using one horned parent reduces the incidence – the kids can be dehorned at birth. The characteristic goat smell comes from glands between the horns and on the neck, which enlarge during the breeding season. The male has the unlovely habit of spraying his head, chest and forelegs with urine and sperm.

The birthweight of kids may vary from 2 to 5 kg, depending on litter size and breed. Single males can grow at 320 g per day, but a more usual figure is 200 g per day.

The goat is a ruminant like the sheep and the cow, but is better able to digest tough

118

fibre. Goats have mobile upper lips and long grasping tongues that enable them to graze very short grass, and also leaves, which are not normally eaten by other domestic livestock. In a free range system they prefer a varied diet, choosing to eat brambles, shrubs, short grass and weeds. They do not do so well on long lush or coarse grass. Lactating goats need plenty of minerals in their diet, particularly calcium and magnesium.

Goats can be kept at 10 per hectare, but a rotational grazing system, or frequent drenching with anthelmintics, is needed because they tend to get infected with internal parasites. Despite their catholic feeding habits, enabling them to thrive where other livestock would starve, goats are susceptible to rain and cold and housing is therefore a necessity in Britain.

Meat Goat meat has special features that make it different from mutton. In sheep the fat is distributed throughout the body, whereas with goats the fat is only found round the abdominal organs. This makes goat meat tougher but leaner. Very little work has been carried out on goat conformation, largely because the goat, being a dual purpose animal, is normally killed at a mature stage. However, with the increase in settlers into Britain who have been more accustomed to kid than lamb there should be an upsurge of interest in milk-fed kids.

There are almost no Angora or Cashmere goats in Britain, but the most valuable skins are made into first quality 'glazed kid' leather and the coarser ones into shoes.

Milking Goats are excellent milk producers and mature animals are capable of lactating for more than two years, if left unmated, with daily yields of over 9 litres – less than a cow, but more than a sheep. The milk has special characteristics which make it more suitable for people who are allergic to cow's milk. It is alkaline, not acid, with thinner, smaller fat globules and the curd is softer. Numerous cheeses are made from goat's milk; some of the most common are pictured below.

Given the milk- and meat-producing qualities of the goat, it is surprising that so few are kept in Britain. They can live on short grass and leaves, and cost a relatively small amount to buy compared with a dairy cow. It is only the wet British climate that is against them; in southern Europe the goat is very popular.

Cabrales

Valençay

Gjetost

Curd cheese

Horses

The heavy horse has declined considerably in numbers since 1945 and is now hardly used at all on British farms. There are signs, however, of a renewed interest in it, partly because, unlike a tractor, it does not need fuel oil, and partly because it can be used in very wet conditions when a tractor would damage the soil. In the past, heavy horses were used

to carry men in armour to war and wherever the Romans built their roads in western Europe heavy horses have traditionally been popular, for they can pull heavy loads over sound surfaces. Just as the tractor replaced the horse because it was quicker and more powerful, so the heavy horse replaced teams of oxen in most parts of Europe.

British breeds The Shire was the great horse of England and was developed to carry armoured knights weighing 200 kg into battle. The Shire of today is larger than its forebears, weighing nearly 1000 kg and averaging 17 hands (1.7 m) high, with heavy bones, heavily feathered (hairy) legs and usually with white 'stockings'. It is bred largely in the Lincoln and Cambridge area, is docile and capable of pulling a net weight of 5 tonnes.

The Clydesdale (800 kg) is slightly smaller than the Shire but is more active; indeed it was bred for this and for the exceptional wearing quality of its feet and limbs. It stands 16.2 hands (1.68 m) high, and can be bay, brown or black with considerable white on the face and legs. The breed dates from the mid-eighteenth century and the Clydesdale was much used in the coalfields of northern England and Scotland.

The Suffolk, or Suffolk Punch, is always chestnut. It weighs 950 kg, stands 16 hands (1.6 m) high and has clean legs (neither feathered nor white). The early Suffolks could, it was said, only walk and draw; they could trot no better than a cow. Gradually, however, they were developed into a handsome, lighter and more active horse. Suffolks often feature in hauling exhibitions and with teams pulling against each other in shows.

Shire
One of the largest
breeds of draught
horse

Clydesdale
Smaller than the
Shire, but with
strong legs and
feet

Suffolk
A breed often seen
in shows or hauling
exhibitions

Foreign breeds The Ardennais is believed to be descended from the draught-horses praised by Julius Caesar, and is very popular in France, Belgium and Sweden. It is a hardy breed, withstanding harsh climates and poor nutrition, and was used during Napoleon's campaign in Russia and in the First World War. There are two kinds, a small one for the mountains which stands 16 hands (1.6 m) high and weighs 600 kg, and a heavy draught type of up to 1000 kg. The Ardennais is large-boned and is usually bay, roan or chestnut.

The Boulonnais, originally bred in France during the Crusades, is now bred in Boulogne, Picardy, Artois, Normandy and parts of Flanders. It is a very heavy draught-horse, standing 16–17 hands (1.6–1.7 m) high and weighing 600–800 kg, with great bone and muscle. It can be black, bay, red or blue roan and dapple grey.

The Percheron is probably the most widely distributed draught-horse in the world. It comes from Perche in France and, like the Shire, was developed for carrying men in armour. The horse stands 17 hands (1.7 m) high and weighs 1000 kg, but is deep chested with a compact body of great depth and large quarters. The Percheron has clean legs and a good action, and is generally docile and easily handled.

Ardennais
Large-boned and
very hardy

Boulonnais
A powerful breed

Percheron
(right) Docile and
very easily handled

Brabant   Schleswig

The Brabant was originally known as the Flanders Horse or Belgian Heavy Draught-horse. It stands between 16 hands (1.6 m) and 17 hands (1.7 m) high and is usually red roan or chestnut. A powerful, active animal with a short back, a deep girth and feathered legs, it has had a considerable influence on other European breeds.

The Schleswig from northern Germany was much used in the Middle Ages when it was the German version of the British Shire and French Percheron, carrying heavily armoured men. The Schleswig Heavy Draught-horse is now a medium-sized, compact animal of placid disposition. Its predominant colour is chestnut.

Mule

Donkey

Donkeys and mules Domesticated probably on the Mediterranean shores of Africa, thousands of years ago, the donkey is used on smallholdings in southern Europe to carry hay, wood and almost anything else that can be heaped upon its back. Varying from 8 hands (80 cm) high to a large type of 11 hands (1.1 m), which is often white and is much prized in the Middle East, it is hardy, patient and stubborn, with a large head and ears and small feet. It prefers dry sunny regions.

The mule, the sterile offspring of a jackass and a mare horse, is also used as a pack animal. Hardier than the horse, it requires more attention than the donkey.

Pigs

The pig is not a ruminant and therefore cannot deal efficiently with fibre, even though its caecum acts as a fermentation vat. Over 10 000 years ago, when man still lived in caves, the pig and the dog were competitors for food, unlike the sheep and the goat which were able to live on grass and leaves. To become a villager, man needed a reliable food source to harvest. When this was provided by growing cereals and there was an excess of grain to human needs the pig was encouraged – although three-quarters of the potential energy of the cereal grain is lost when cycled through the pig.

The pig in ancient Britain lived mostly in woodlands and was an important source of meat to the Romans, the Saxons and the Normans. By the Tudor period large numbers were kept in sties but were still turned out to woodland and waste ground where they could forage for earthworms, acorns and bulbs.

As with other farm livestock, the eighteenth century witnessed the beginning of selective breeding. The Large White was fixed as a clearly defined breed, and between 1770 and 1780 Chinese pigs were imported for their earlier maturity and their meatier carcasses.

By 1850 pig shows had become popular and a start was made on classifying different breeds; this in turn led to the formation of Breed Societies. During the Second World War and its aftermath, however, there was so much emphasis on weight alone that meat quality began to decline and it was not until 1953 that this aspect was re-emphasized by the adoption of a carcass grading scheme.

Today 14 million young pigs are slaughtered each year in Britain and the national pig herd contains nearly 8 million breeding sows. Denmark has over 9 million, Italy, the Netherlands and Spain nearly 10 million each, France and East Germany nearly 12 million each and West Germany has over 22 million; hence the universal German wurst. In Britain there is an increasing number of large herds and the average herd size (breeding and feeding pigs together) is 206 pigs, which is the largest in the EEC; the Netherlands comes next with an average of 127, and then Denmark with 84. Each year 19 kg of pork, bacon and ham are eaten per head of Britain's population, which is slightly less than the amount of beef and veal that is eaten but more than the consumption of poultry, and of mutton and lamb.

Early-maturing pigs, porkers, reach the desired carcass conformation for pork at 45–64 kg live weight. Later-maturing pigs, baconers, do the same at 90–100 kg live weight, while the heavy hogs are slaughtered for manufacturing products such as pies, sausages and tinned hams at 118–128 kg. All slaughtered pigs, at whatever weight, should have the maximum amount of lean meat and a minimum of bone and fat.

Shows Pig shows are popular today and showing has become an art in itself, with all the anointing, shampooing and grooming that goes with it. The problem with showing pigs in the ring, however, is the lack of direct control over the animal, particularly with a boar. With no restraint such as a head collar or harness, the handler is restricted to a stick and a board. If a boar should choose to wander out of the ring, or to attack a rival at speed, a light stick is only a mild deterrent.

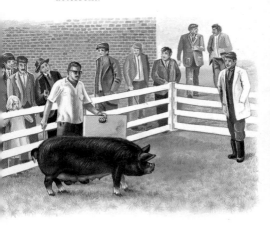

British breeds The Large White has been called the universal breed because it is found in many countries throughout the world. It is the most numerous breed in Britain and can be used for pork, bacon or the heavy hog trade. The sow averages 10.6 pigs born and 9 reared per litter, with a farrowing interval of 187 days; this means 18 piglets reared per sow in 374 days. Large Whites have a high lean content, excellent food conversion and a high daily gain (802 g per day for baconers). Mature boars may weigh up to 510 kg.

The Welsh pig is long and white with lop ears. It is hard to distinguish from the British Landrace and renewed interest in it was stimulated by the importation of the Swedish Landrace. The Welsh pig is prolific, with a short farrowing interval, and is primarily a baconer.

The British Landrace developed from the importation of the Swedish Landrace in the early 1950s. It is the longest breed and produces high quality porkers and baconers. It has a better growth rate and shows more efficient food conversion than the Large White, producing extremely thin backfat, but is not as good for muscle quality.

Hybrids are commercially important, and the Large White × Welsh or Large White × British Landrace are popular crosses.

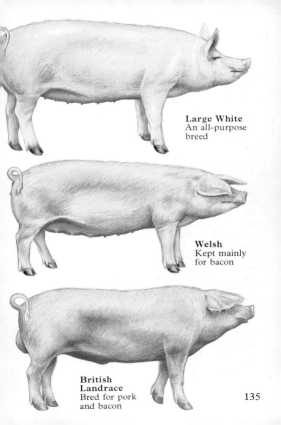

Large White
An all-purpose
breed

Welsh
Kept mainly
for bacon

**British
Landrace**
Bred for pork
and bacon

135

Apart from the three already mentioned, there are only a further six British breeds of pig.

The British Saddleback was formed by amalgamating the Essex and Wessex Saddlebacks in 1967, previously treated as separate breeds. It is a hardy, prolific animal suitable for outdoor rearing and is used on the white breeds to produce first-cross porkers, baconers and heavy pigs.

The Tamworth resembles the ancestral wild pig and is well-suited to poorer food and the outdoor life. In Britain it is mainly used for pork and bacon production.

The Large Black is a heavy pig (500 kg at maturity) with long lop ears reaching to the end of the snout. The hardy sows are docile and are useful for crossbred pig production.

The Berkshire has a short head and body and prick ears. It is almost black in colour, with white feet and white on the end of its tail and snout.

The Middle White was produced by crossing the Large White and a now extinct breed, the Small White. Like the Berkshire, it is early maturing and useful for pork production.

The Gloucester Old Spot is a hardy outdoor animal, known as the 'orchard pig' because it originally lived on windfalls in the Severn Valley.

British Saddleback

Tamworth
A rare but hardy breed

Large Black
Used to cross into other breeds

137

Foreign breeds The Pietrain from Belgium has a high proportion of lean meat and is double-muscled, or extra-muscled, particularly in the hams. It does, however, produce a proportion of poor quality or 'PSE' carcasses (pale, soft and exudative).

Of the European Landraces only two have been chosen here, because they represent opposite extremes. Landrace means 'the local, native or indigenous breed'.

Denmark has the best-known breed of Landrace, which is used as a specialist bacon pig mainly for the British market. The Danish Landrace is therefore long, with extremely thin backfat, and with average ham shape and loin eye area.

The Belgian Landrace has well-shaped hams and bulging loins and has been bred for the fresh and manufactured meat market. Of the Landrace breeds it scores highest for loin eye area, killing out percentage, lean content and ham percentage. Like other muscular breeds, however, it is susceptible to stress, is rather unsound on its feet, and has a proportion of pale and often PSE carcasses.

Other foreign breeds found in Britain are the prick-eared Hampshire from North America, which is long, lean and black with a white saddle, and the red Duroc, which is popular world-wide as a producer of fine lean meat.

Pietrain
Noted for its
lean meat

**Danish
Landrace**
Mainly bred
for bacon

**Belgian
Landrace**
Kept for pork
and ham

139

Pig-keeping indoors This has become a carefully managed enterprise, so allowing predictable rates of growth. Controlled ventilation is used to prevent temperature variation, which causes fluctuation in growth rate and, below a critical temperature, means that the animal has to eat extra food to maintain its body temperature. Pigs are comfortable at humidity levels over 50 per cent, but low temperature and high humidity lead to extreme discomfort and poor performance. In addition, scrupulous precautions have to be taken to avoid the build-up of such infectious diseases as enzootic pneumonia.

Pig-keeping outdoors A huge field with a large number of sows, each with her ark and litter of piglets, lying contentedly or rooting about, makes an entertaining sight on a fine summer day. Outdoor systems can and do · produce healthy, fast-growing weaners, but if conditions are wet the ground becomes a quagmire and pig movement, collection and supervision are difficult. Although mucking out is reduced, maintenance and labour costs are high and the ground has to be free-draining and not 'pig-sick'. Coloured breeds, such as the Tamworth, are more suited to outdoor pig-keeping than the white ones.

141

The boar A boar is sexually mature at 5–6 months of age, at a weight of some 100 kg, but should be used sparingly until a year old. He must be kept in peak condition, because an overweight boar is frequently unable to mate successfully.

A young boar should be started with a gilt, or young sow, who is on heat for the first time – a mature sow is likely to bully him. If ready to mate the sow will signal her willingness by standing rigid when she smells the boar, or rather the steroids in his white frothy saliva. If she is not yet ready to stand then the experienced boar will pursue her, nosing her flanks and uttering a mating song of soft gutteral grunts.

A serving crate – a box with slatted sides and bars at either end – aids mismatched matings. The sow is placed inside it and if the boar is much larger than the sow he can rest his legs, and weight, on the top of the crate. If he is smaller then a platform can be built for his hindlegs.

Artificial insemination With artificial insemination (AI) semen from a good boar can be used over many sows, so increasing the rate of genetic improvement. It also makes the spread of disease less likely and relieves the farmer of the need to keep stud boars.

Boar semen is diluted after collection by

about one-tenth (compared with one-thousandth for bull semen) and can be kept chilled for three days or deep frozen as a pellet for longer. Thawed sperm deteriorates rapidly outside the body and so must be used promptly.

Conception rates with artificial insemination are not so good as with natural service. The pig ovulates during oestrus, which is brief and difficult to identify, and insemination must be carried out then.

The boar entices the sow with a mating song

Reproduction A sow can breed all the year round, which means that pig breeding can be very accurately planned. Oestrus lasts for just over two days and the oestrous cycle for 21 days. Pregnancy lasts for 115 days, and so if the piglets are suckled for 56 days and the sow served, now in oestrus again, within five days of weaning, then the whole cycle takes 176 days (115 + 56 + 5) and two litters can be produced per year. Most sows are kept for six litters, or for two and a half years after first breeding, because after this the number of live piglets per litter declines. A gilt, a young sow, is ready to breed at six months of age, when she should weigh about 100 kg.

One to three days before farrowing, the sow will start to build a nest with straw or whatever is available. Farrowing usually starts at night, with one piglet being born about every ten minutes. It is best done in a farrowing crate (pictured opposite), which will prevent the sow from lying on her piglets or from eating them. Piglets only weigh 1.5 kg at birth and are born with sharp little teeth which should be removed to prevent teat damage.

Within a day or two of birth each piglet has its own teat, with the biggest at the head end, where there is most milk, and the runts towards the tail.

Nutrition and management Pigs consume a very wide variety of foodstuffs, ranging from acorns, roots and grass in the wild to edible domestic waste, brewing by-products and skimmed milk indoors. Pig food may account for three-quarters of the total running costs of a pig unit, and is either bought in ready mixed from a feed firm or is mixed on the farm. A large arable farm will usually produce its own pig food.

Feed can either be fed 'wet' on the floor, particularly if pelleted, or 'dry' in troughs. Experiments to compare wet and dry feeding of mash show that wet feeding increases growth rate and decreases the feed-to-gain ratio. Mechanized pipeline feeding was introduced in 1950.

Growing pigs can be fed to appetite up to about 40 kg live weight, but after that the level of feed should be restricted to prevent excess fat being laid down. Although free feeding gives the fastest growth rate, it is at the expense of food conversion efficiency. Pregnant sows are normally fed 1.8–2.3 kg of a 15–16 per cent protein ration per day, but during lactation they can be fed to appetite (4.5 kg of dry matter per day). Young piglets also need copper, iron to prevent anaemia, and a palatable feed in pellet or crumb form, especially if weaned early. The earlier piglets are weaned (as young as seven days old), the more litters and piglets the sow will produce per year.

The pig manager should be scrupulous in his routine, dealing with his pigs quietly and kindly, because few animals are so amenable to training. Weaners can be graded and matched for size to minimize fighting and bullying, as a runt or sick pig is soon victimized. It helps to spray them with pig oil at this time to disguise litter identity. Growing pigs do best in conditions that are as tranquil as possible. A full pen is defined as one where each animal should have enough room, but no more, to lie down when sleeping.

Meat When a pig is slaughtered, two things are done promptly: as much blood as possible is removed and the viscera (heart, lungs, stomach and intestines) are taken out. This reduces the number of putrefying bacteria and removes their food source. The carcass comprises the body less the internal organs and a few other parts. The head, skin and tail are left on, whereas they are removed from the carcasses of cattle and sheep, so that the producer is paid for 70–75 per cent of the live weight and not for 40–55 per cent as in cattle and sheep.

Quality is the proportion of fat to lean meat, the relationship of these tissues and their tenderness, succulence, colour and taste. The butcher is concerned with the proportion of bone and gristle to edible meat, the shape of the portions for domestic cooking and their appearance. He does not want to buy waste and bone. As a pig grows older, its muscles contain less water and more fat. A good carcass should have firm fat, for fat and lean are not well intermingled in pig meat and soft fat is no use for curing.

The pork cuts are shown opposite. The leg, loin, and neck are usually sold for roasting, while the belly and fore-end are pickled in salt and then used for boiling. 'Crackling' is the natural skin on the outside of a roasted joint.

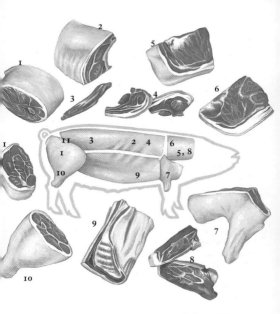

Pork cuts 1 Fillet half leg, 2 Loin, 3 Fillet,
4 Loin chops, 5 Spare rib, 6 Blade, 7 Hand and
spring, 8 Spare rib chops, 9 Belly, 10 Knuckle,
11 Chump chops

Bacon and ham Texture and colour are important selling features of these products and nitrite is used to give a strong pink colour to some hams. Tasting panels have found that palatability increases in proportion to the amount of fat present, until it forms about 38 per cent of the meat. Nitrogenous extractives, the end products of protein metabolism, produce the distinctive flavour of meat. The flesh of birds and most game is particularly rich in them.

Shortly after death, alterations in the chemical make-up of a carcass begin to affect the texture of its meat. The glycogen changes to lactic acid, which permits the salt in curing to penetrate rapidly. The longer meat is hung in cold storage the more tender it becomes.

In the bacon pig an immature carcass of less than 64 kg will have too low a proportion of lean and fat to bone and the 'streaky' cuts may not be sufficiently thick, while one weighing 82 kg will have too thick a layer of fat over all the cuts. The bacon cuts are shown opposite.

Pig meat was originally cured to preserve it, just as pemmican and biltong are produced by salting and drying. In the older, Wiltshire cure, the side was injected with a strong brine of salt, sodium nitrate, saltpetre and (sometimes) sugar, and then stacked with others in a deep tank and

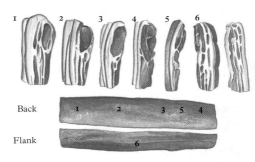

Back

Flank

Bacon cuts 1 Top back, 2 Middle cut, 3 Short back, 4 Long back, 5 Oyster cut, 6 Streaky

covered with more brine for 4–5 days. The tank was then drained and the sides re-stacked to mature before being dispatched as green bacon or after smoking. The more recent mild cure is much quicker. The carcasses are chilled, cut into four major joints, trimmed and boned and then injected with a special brine. Cured bacon can be smoked over smouldering oak sawdust or chips for 12 hours; 'country cures' are smoked for longer. Heavy pigs are used for processing into tinned meats, sausages, pies, etc., but their carcasses also provide bacon and pork cuts.

Poultry

Chickens Chickens provide nearly all the eggs and most of the poultry that we eat. All breeds are believed to have descended from the Jungle Fowl (*Gallus gallus*), a small bird of Asian tropical forests. Domestic chickens were kept in China as long ago as 2000 BC, reached some of the Mediterranean countries by 300 BC and spread to western Europe and Britain by AD 100. The oldest British breeds are the Dorking, probably introduced by the Romans, and the Old English Game, developed as a fighting cock.

Chicken breeds can be divided into light and heavy, or classed by their region of origin, but a more practical classification is into laying, table, dual-purpose and fancy breeds. Laying breeds, which emphasize egg production, laying mostly white-shelled eggs, mainly in the spring and summer, include the Leghorn, Welsummer (brown eggs), Recap, Ancona, Andalusian, Minorcas and Hamburgh. Table breeds such as the Dorking and Faverolle, kept for meat production, are larger and deeper-bodied than the layers. Dual-purpose breeds are intermediate types combining the qualities of both egg layers and table birds. Their eggs, usually tinted or brown, are mainly produced during the winter. These are the typical farmyard birds and include the

Rhode Island Red

White Leghorn

Light Sussex

Black Leghorn

Rhode Island Red, Sussex, Plymouth Rock, Wyandotte, Cochin, Orpington, Australorp, Barnevelder, Brahma, and Marans. Not all chicken eggs are white or brownish shelled – those of the South American Aracauna breed are blue. Fancy breeds, kept mainly for their appearance, include all game bird breeds and bantams. Bantams are true-breeding miniatures of the larger breeds, about a quarter of their weight. Within a breed there is usually a number of varieties and sub-varieties. White, for example, is a variety of Leghorn and the Rose-combed White is a sub-variety of the White Leghorn.

First crosses, which are reputedly hardier than pure breeds, are produced by mating two pure breeds. A well-known first cross is obtained by mating a Rhode Island Red cock with a Light Sussex hen. In this cross colour is sex-linked, enabling chicks to be sorted easily into males and females. Recognition of sex-linked factors has led to the creation of a number of new breeds since the 1920s which show sex distinction on hatching. Each was created using the Barred Plymouth Rock and another breed. Most chickens these days are commercial hybrids, produced from several breeds. Hybrids started in the 1950s and became increasingly popular as poultry production became more and more intensive.

Buff Plymouth Rock

New Hampshire Red

Brown Leghorn

Silver-laced Wyandotte

White Cochin

Dark Brahma

155

Black
Orpington

Welsummer

Silver-grey
Dorking

Redcap

Australorp

Old
English
Game

156

Faverolle

Dark
Cuckoo
Marans

Ancona

Andalusian

Black
Minorca

Silver-spangled
Hamburgh

157

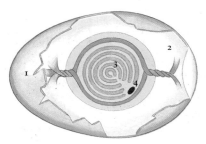

Structure of the egg

The chicken's egg About 11 per cent of the egg is shell (**1**), 60 per cent white (or albumen) (**2**) and 29 per cent yolk (**3**). The albumen consists of four layers in alternate dense and liquid forms. The yolk too is a series of lighter and darker concentric layers. On average 12.1 per cent of an egg is protein, 10.5 per cent is fat, 65.6 per cent is water and the remaining 11.8 per cent is made up of carbohydrate, minerals and vitamins. A typical egg has an energy value of about 85 calories.

Unfertilized eggs, laid by hens who have not been mated, are produced for eating. They differ only in the germinal disc (**4**) from which the chick develops if fertilized.

Chick development In natural incubation a broody hen sits on fertilized eggs until they hatch but 1000 years ago the Chinese artificially hatched chicks. Today an oven-like apparatus with temperature and humidity controls is used. Eggs in an incubator should be turned at least three times a day. After 21 days the chick hatches out by pecking through the shell.

The extra warmth chicks need for the first six weeks of their lives is provided by a box with an infra-red lamp. Females start to lay eggs at 4–5 months old and continue for about a year. After a rest period, called a moult, they start again. From beginning to lay until her first moult a hen is called a pullet. A young cock is called a cockerel.

15 days

21 days

Feeding and digestion Chickens have a simple digestive system and eat a mixed diet. Unlike ruminants, they cannot deal effectively with bulky or fibrous food. Traditionally, farmyard hens were fed on kitchen scraps and spare corn. Chickens nowadays are normally fed a compound feed, in meal or pelleted form, made up of corn, animal and vegetable protein, minerals and vitamins. A typical hen eats 100–150 g dry weight of food a day. In addition, they should be well supplied with clean drinking water and grit.

Chickens have no teeth. They peck their food and swallow it down the gullet (1) into the crop (2), where it is softened and often stored for several hours before moving to the glandular stomach (3), where digestion begins. It then enters the gizzard (4), where it is crushed and ground before passing to the small intestine (5). The gizzard is a strong muscular organ which needs grit to function effectively. It is colloquially referred to as 'the bird's back teeth'. In the small intestine further digestion takes place and nutrients are absorbed across the intestinal wall into the blood. Undigested food then passes along the large intestine (6), which is headed by a pair of caeca (7), where some further digestion takes place, and is eventually ejected through the anus (8).

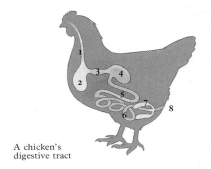

A chicken's
digestive tract

Health and disease A flock can be kept
healthy by proper attention to hygiene,
feeding and general care. Lice and mites are
controlled by regularly dusting the birds
and their living quarters. Many serious
bacterial and viral diseases can be avoided
by preventative inoculation. The build up
of harmful parasites in chicken runs can be
prevented by annually rotating the ground
on which the run stands. Nevertheless,
illness cannot be totally avoided. Unhealthy
or unproductive stock are usually culled,
traditionally by breaking the bird's neck by
hand. The bird should be held upside down
by the legs in the left hand while with the
right its head is pushed down, twisting
sharply to the right. Only healthy birds
should be subsequently eaten.

Egg production There are three basic systems of egg production: battery, deep litter and free range. The battery system is the one used commercially and accounts for over 90 per cent of eggs laid. The hens are kept indoors in individual cages within a controlled environment. Up to 30 000 birds can be kept in one house. When an egg is laid it rolls straight into a collection trough. Food and water are available on demand and droppings fall straight through the floors of the cages. The birds, which are usually hybrids, are put into battery cages at about 4–5 months of age. They are kept there for up to ten months, the most productive part of their lives. During this period they lay on average about 250 eggs, some birds many more. Undoubtedly this is the most productive and efficient system, but it is also the least humane.

The deep litter system is less intensive but more humane as the birds can walk around, though they are not allowed outside. They are housed under cover on deep litter (straw or wood shavings several centimetres deep) with electric lighting overhead. Clean water, special feeding hoppers, perches and nestboxes are provided. The hens are normally replaced by point-of-lay pullets after about a year, when their first

Battery system

Deep litter system

Free range

laying period comes to an end. At this time, the litter build-up is removed and the hen house completely disinfected.

The free range system is the old-fashioned way of keeping hens on the farm. It is the least efficient but most natural system, as the birds can wander about outside and peck, scratch and take dust baths. The chickens are allowed access to a piece of ground during the day and are shut up in the hen house at night, to guard against the cold and predators such as foxes. The hen houses are provided with communal nestboxes – though the hens do not always use them, sometimes preferring more out-of-the-way places! Free range hens are often kept for several years.

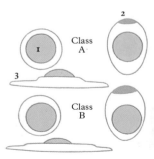

Class A eggs must have a central yolk (1), an air cell of not more than 6 mm (2), and an outer layer of thin white (3). Class B eggs lack one or more of these

Quality and size Eggs are marketed in three quality classes: A, B and C. Class A are fresh eggs which are naturally clean, with the shell intact and the inside perfect. The air cell (the pocket of air inside the egg) must not exceed 6 mm. Class B are eggs downgraded for reasons such as staleness, internal spots or because the eggs have been preserved in some way. The air cell must not exceed 9 mm. Class C eggs are those fit only for breaking-out purposes, they cannot be sold to the public. Some egg boxes carry a red band labelled 'extra'. These are super-quality eggs which have been packed for less than a week and have an air cell not exceeding 4 mm.

Eggs are graded for size according to weight. Size 1 is for eggs of 70 g or over; size 2 weigh 65 g–under 70 g; size 3, 60 g–under 65 g and so on. The smallest size are 7 which weigh under 45 g. Each size category carries a colour code on the egg box: brown is 1, blue is 2, purple is 3, red is 4, green is 5, black is 6 and 7 is white.

Colour codes

Meat production The production of chickens for the table may be broadly classified into broilers and capons. Broilers are reared under contract in large flocks, usually in excess of 20 000, and sold off the farm for slaughtering and processing in big, modern packing stations. Production is very specialized, requiring fast growth and a high rate of conversion of food to meat. Broilers are usually marketed when they are 8–11 weeks old, on reaching a live weight of around 1.5–2.5 kg, and they provide the frozen chickens that we buy at the supermarket. The birds are special, white-feathered hybrids which have a plump and uniform shape. Birds finished younger than this, at about 5–7 weeks, are known as poussins. One serving usually consists of an entire bird. Broiler strains or cockerels from laying strains are used for poussin production.

Capons are cockerels which have been surgically or chemically caponized (emasculated) and specially fattened. They are usually slaughtered at 12–15 weeks of age at about 3–4 kg live weight, and often marketed as fresh chickens. They tend to be reared in smaller flocks on general farms. Capons are noted for their tenderness and succulent taste and have a large proportion of white meat to dark.

Until World War II, spent laying hens

made a significant contribution to the supply of table poultry. Nowadays finished hens normally end up as ingredients for processed foods such as soups, pastes and pet food, though there is a market for them as stewing or boiling fowl.

Drugs play an important role in broiler production. They are used in disease prevention and control and for growth promotion. Without drugs to prevent diseases modern broiler production, with its total reliance on intensive methods, would be uneconomic. Growth-promoting drugs are used in nearly all broiler feeds as they improve growth by increasing the efficiency of food conversion.

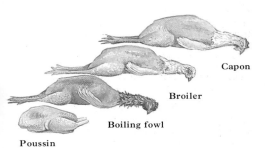

Capon

Broiler

Boiling fowl

Poussin

Turkeys The wild turkey is native to North America. It was taken to Europe in 1500 by Spanish explorers of Central America and reached England from Spain about a quarter of a century later. In Britain, breeds range from black (Norfolk Black, developed in East Anglia in the nineteenth century) and blue (Blue) through brown (Buff) to white (Beltsville and British White). The Bronze was the major British breed for most of this century until commercial breeders changed to hybrids. In this and some of the larger breeds selective breeding, emphasizing large breasts and short legs, has made natural mating virtually impossible. In the late 1960s commercial producers changed over to white hybrids and today table turkey production is an important industry run along similar lines to that of broiler chickens. It is continuous through the year, peaking in the weeks leading up to Christmas. Turkeys are more difficult to rear than chickens as the young poults quickly succumb to chills if the temperature of their environment is not correctly maintained. Birds are marketed in three categories: mini turkeys, slaughtered at 12 weeks and weighing about 4 kg, midis, slaughtered at 16–17 weeks and up to 7 kg, and maxis, slaughtered at 24 weeks and around 14 kg.

Turkey Breeds

Bronze

British White

Norfolk Black

169

Duck Breeds

Aylesbury   Pekin

Ducks Domestic ducks originate from the wild Mallard (except for the Muscovy, descended from the South American Musk Duck). Much less common on farms than they used to be, they are traditionally kept on free range, with a duck house and access to a pond. Good scavengers of insects, slugs, water fleas and grass, they are fed a diet similar to chickens. The main commercial production for the table uses white hybrid broilers based on the Pekin breed, which has now largely replaced the Aylesbury.

Duck keeping for eggs has declined but duck eggs are larger than hen's eggs and have a slightly stronger taste. All domestic breeds have an incubation period of 28 days, except the Muscovy which needs 35 days. Offspring produced by crossing Muscovies with any other domestic breed are sterile.

Magpie

Khaki Campbell

Muscovy

Indian Runner

Rouen

Buff Orpington

171

Geese Descended from the wild Greylag, geese have been domesticated since ancient times – Egyptians, Greeks, Romans and Chinese all kept them. They are noisy creatures and famed watchdogs. Reared chiefly as table poultry, traditionally for Christmas, they are also kept to provide liver for pâté, for their feathers and for their eggs. Geese are predominantly grazers and foragers so their production is usually linked with grass growth, making table birds generally available in autumn and early winter. When the grass supply is insufficient it is supplemented with grain. Geese have not been subjected to intensive farming methods and usually are kept in small flocks and allowed to range with only basic shelter and access to grazing. They spend far more time on land than do ducks and do not need a pond.

Breeds are classed as domestic or ornamental. Most famous are two heavy breeds, the Embden and the Toulouse. A cross between them is widely used for meat production in Britain. The Toulouse originated in France where it is bred for the production of pâté de foie gras. The Brecon Buff is a medium-size breed founded on stock from Breconshire hill farms. The Chinese and the Roman are the most popular of the smaller breeds in Britain.

Goose Breeds

Chinese

Toulouse

Embden

Brecon Buff

Grassland

Grass is one of the major agricultural crops of the world and grasslands and grazing lands cover almost a quarter of the world's land surface. In Britain this proportion is even higher – just over two-fifths of the country is under grass. These grasslands play the dominant role in feeding the nation's 45 million ruminant livestock.

A grass sward is made up of many individual shoots called tillers. In this country, the grass growth season starts in early spring and ends in late autumn. Growth is rapid in early spring, but as the days get longer and warmer, the tillers begin to produce flower heads which eventually seed if the crop is not cut or grazed. A bigger crop of higher quality is obtained from a sward over the year if the grass is not allowed to flower and produce seeds. A sward managed in this way is most productive in the spring, and its second most productive period is in the late summer and autumn.

British grasslands vary greatly in their botanical composition and contain not only grasses but other plants as well, including clovers, herbs and various weeds. They can be broadly classified as uncultivated or cultivated grassland; the cultivated category is usually further subdivided into permanent and temporary grassland.

Sheep's Fescue
Festuca ovina

Bilberry
Vaccinium myrtillus

Uncultivated grassland In Britain over 2.5 million hectares of rough grazings occupy mostly hilly and mountainous areas and include a wide variety of vegetation. Of generally poor quality, they usually support hill-sheep farming. Poorest grazings are the very poor soils of mountainous areas over 600 m, where vegetation consists mainly of mosses, lichens and occasional areas of Sheep's Fescue and bilberry. Moors and heaths are more important, where Deer-grass and Cotton-grass predominate in boggy areas and heather in dry areas. There are over a million hectares of heather in Britain, a useful food source for sheep and rich in minerals. The crop is improved by burning every 10–20 years, which helps maintain it at a high nutritive value. The best rough grazings are the *Agrostis-Festuca* dominated pastures of the lower hill slopes. In other upland areas they are dominated by Purple Moor-grass and Mat-grass.

Plants of moor and heath: **1** Cotton-grass *Eriophorum angustifolium*, **2** Mat-grass *Nardus stricta* (on wetter sites where Sheep's Fescue common on drier ones), **3** Deer-grass *Scirpus cespitosus*, **4** Heather *Calluna vulgaris*

Valuable grasses common on lower hill-slope
rough grazing include: **1** Common Bent
Agrostis tenuis, **2** Brown Bent *Agrostis canina*,
4 Red Fescue *Festuca rubra*. **3** Purple
Moor-grass *Molinia caerulea* is of limited
nutritional value

Rough grazing on most seashores commonly
includes: **1** Common Saltmarsh-grass
Puccinellia maritima, **2** Marram-grass
Ammophila arenaria, **3** Sea Couch *Agropyron
pungens*, **4** Common Cord-grass *Spartina anglica*

Cultivated Grassland Lowland grasslands support intensive dairy, beef and sheep production. Most of Britain's 7 million hectares of cultivated grassland is permanent pasture, found mainly in the higher rainfall areas of the north and west. This is cultivated grassland sown for an indefinite period, usually for five years or more. Temporary grassland is sown for a specific period, usually as part of an arable rotation to build up soil fertility and clean the land. Temporary grass, termed a ley, is normally less than five years old.

The productivity of permanent grassland depends upon its botanical content. High-yielding permanent pastures tend to be dominated by Perennial Rye-grass, the most commonly sown grass species in British agriculture, which is very responsive to nitrogen fertilizer, provides good grazing, is hard wearing and makes good hay and silage.

Clovers and other legumes, such as trefoil and vetch, are often found in permanent grassland. They are important because they can fix atmospheric nitrogen in their root nodules, thus making a valuable contribution to the nitrogen supply of the sward. White Clover is used in many grass seed mixtures and certain herbs are also sometimes added. Chicory, Hoary Plantain and Yarrow are all herb species rich in minerals and palatable to livestock which grow naturally in old pastures.

Temporary grass is sown as either a short or a long ley. Short leys are down for one or two years, whereas long leys are down for three or more years. Italian Rye-grass or Perennial Rye-grass are the usual grass for a

Legumes of permanent grassland include:
1 White Clover *Trifolium repens*, 2 Hop Trefoil *Trifolium campestre*, 3 Common Vetch *Vicia angustifolia*

short ley, sometimes sown together with Red Clover. Long leys are usually based on Perennial Rye-grass and often include some White Clover.

Lucerne (also called alfalfa) is a deep-rooted legume which is highly resistant to drought and is usually grown as a crop on its

Herbs rich in minerals which grow naturally in old pastures and are sometimes added to seed mixtures: **1** Chicory *Cichorium intybus*, **2** Hoary Plantain *Plantago media*, **3** Yarrow *Achillea millefolium*

own, though sometimes with a little grass. It is normally cut before being fed to the animals as it is not particularly well-suited to grazing. Sainfoin grows similarly to lucerne but requires chalk or limestone soils. It is very suitable for hay or silage making and good-quality sainfoin hay is used for racehorses.

Short leys often include: 1 Red Clover *Trifolium pratense* with 4 Italian Rye-grass *Lolium multiflorum*. 2 Lucerne *Medicago sativa* and 3 Sainfoin *Onobrychis viciifolia* are usually grown alone

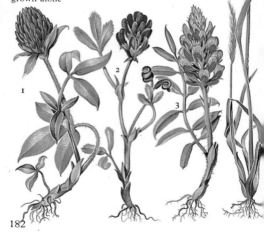

Important species in permanent grassland:
1 Perennial Rye-grass *Lolium perenne*,
2 Cocksfoot *Dactylis glomerata*, **3** Meadow
Fescue *Festuca pratensis*, **4** Timothy *Phleum
pratense*

Some less important grasses of value to
livestock which are found in well-established
pastures but seldom sown nowadays include:
1 Crested Dogstail *Cynosurus cristatus*,
2 Common Foxtail *Alopecurus pratensis*, **3** Sweet
Vernal-grass *Anthoxanthum odoratum*, **4** Tall
Fescue *Festuca arundinacea*, **5** Annual Meadow-
grass *Poa annua*, **6** Rough Meadow-grass *Poa
trivialis*, **7** Smooth Meadow-grass *Poa pratensis*

5

6

7

Many weed species are found in cultivated grassland, especially in old pasture. Some of the commonest are: **1** Yorkshire Fog *Holcus lanatus*, **2** Couch (or Twitch) *Agropyron repens*, **3** Barren Brome *Bromus sterilis*, **4** Soft Brome *Bromus mollis*, **5** Creeping Bent *Agrostis stolonifera*, **6** Black Bent *Agrostis gigantea*, **7** False Oat-grass *Arrhenatherum elatius*

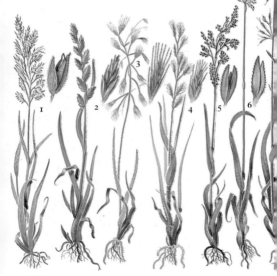

Management of cultivated grassland

Good grassland needs careful management if it is to remain productive. Correct liming and manuring, and adequate drainage, are very important, together with good grazing management. Lime reduces soil acidity. Very acid soils discourage the growth of the better grasses and clover, and inhibit the action of manure and fertilizers. As a result weed species such as bents and Yorkshire Fog take over.

Adequate land drainage is essential for profitable grass production. Poorly drained soils reduce grass growth, encourage weeds, such as rushes and buttercups, and animal and plant diseases. Livestock and heavy machinery can cause serious structural damage to such soils. The trampling of grassland by livestock (poaching) is a major limitation on grass growth. Most acute during winter grazing on low-lying, poorly drained soils, it can also result from over-stocking, even on well-drained land.

Ditches and under-drainage are the two basic types of artificial drainage. Ditching, the old way of draining land, involves digging open ditches on the edge of the field – now sometimes replaced by piped drains which are easier to maintain. Under-drainage, the modern method, is a network of artificial channels below the surface which may be formed of clayware pipes, called tiles (tile-drainage), or cut by a special plough (mole-drainage), a cheap and effective way of improving heavy soils. The blade of the mole plough has a bullet-shaped end which cuts a drainage channel in the subsoil clay. It also makes fissures in the subsoil, helping to break it up and aid drainage. In the right soil, mole drainage can be effective for 5–10 years. In practice, under-drainage often consists of tile- and mole-drainage used together.

To grow well, grass needs a well maintained supply of essential nutrients, such as nitrogen (the most important), phosphate and potash. Most of them are supplied by

Manuring with a side-spreader

Poaching is worst near gates and food
and drinking troughs

manures and fertilizers. Manures are bulky
organic substances derived from farm or
waste products, the most common being
farmyard manure consisting of dung, urine
and various litters used in bedding. A cow
produces 3–4 tonnes of manure a year.
Slurry, a liquid mixture of dung, urine and
washing water or rain, is another common
manure. Intensive systems of keeping cattle
and pigs, using less bedding, have increased
slurry production. Manures supply plant
nutrients and add organic matter to the soil,
improving its condition. Fertilizers, normally inorganic, water-soluble salts with a
high concentration of plant nutrients, are
usually applied to grassland as small dry
pellets.

Grazing If a sward is to be productive, the grass must be regularly and effectively harvested throughout the growing season. Different types of livestock graze in different ways. Sheep can eat short grass and bare their teeth to graze. They can graze right down to ground level. Cattle need longer grass, as they graze by wrapping their tongues around a clump of grass and chopping it with their teeth. Pigs are not too fond of grass, they prefer clovers and herbs. All grazing animals prefer young, leafy material. Sheep can be more selective than cattle, as the grazing action and larger jaws of cattle do not allow such precise selection.

For animals at pasture, the number of livestock a piece of grassland can support is influenced by the farmer's choice of grazing system. There are six basic systems, each with its advantages and disadvantages. The appropriate choice depends upon the particular circumstances:

1 Zero Grazing. Animals are kept in yards or buildings throughout the year and fed grass cut and carted to them during the summer and hay or silage the rest of the year

2 Free grazing. Unrestricted access to a large area so that animals eat where and what they want. It makes poor use of grassland but is often the only method practicable on land such as rough grazings.

3 Set Stocking (continuous grazing). A number of animals are turned on to a field and remain there for much of the grazing season. Fencing and supervision costs are less than with other forms of control

4 Paddock grazing (rotational grazing). A field is subdivided into paddocks (usually not more than eight) and each is grazed in turn for a few days by a whole flock or herd and then rested

191

5 Creep grazing. A refinement of paddock grazing. Creep gates in paddock fences allow lambs or calves to feed ahead of other animals. It allows the young access to the best grass without competition from older animals

6 Folding (strip grazing). The best use of grazing: a fresh supply of grass or forage crops is provided once or twice a day by moving an electric fence. A back fence is often also used

Haymaking Haymaking is the oldest and most common method of conserving grass for winter feed. Good quality hay is green and sweet smelling. It is made by cutting grass with a mower, preferably in May or June before the grass flowers and gets stemmy, and leaving it to dry naturally in the field. The drying process is speeded up by tedding, which tosses and turns the grass. When the hay is dry enough it is put into swaths and baled for storage. The moisture content should be down to 5–20 per cent before it is stacked. There are two types of bale – the small, rectangular bale weighing about 25 kg and the large bale weighing some 500 kg. Large bales are either cylindrical or rectangular and make

Tedding with a spring-tine hay turner. This tractor and one on page 223 contravene safety regulations by not having a cab and would not now be permitted on British farms.

handling from field to store a relatively speedy task given the right equipment. Bales are usually stored in a Dutch barn, though big bales are often stored in the open. A Dutch barn is a basic storage building. It usually consists solely of a roof on a frame, though sometimes one side or more is cladded.

Fine weather is essential for making good hay. In wetter areas the drying process is sometimes speeded up by putting the partly-dried grass on to tripods or even over fences to dry. The modern method of speeding up the process to reduce the risk of bad weather is barn drying. Partly-dried grass is brought indoors and currents of hot or cold air passed through it.

Baling

Ensiling Ensiling is a method of conserving grass or other green stuff for winter use by pickling it. The product of ensiling is silage and the container in which it is made is a silo.

Ensiling is a fermentation process in which carbohydrates within the plant are converted into lactic, butyric and acetic acid by bacteria carried on the plant. The acids produced preserve the grass. In well-fermented silage lactic acid production must be encouraged at the expense of butyric acid. Air must be excluded and excess water allowed to drain away. The fermentation can be improved by additives such as molasses or formic acid. Well-fermented silage is bright, usually yellowish brown, with a pleasant acid smell. Butyric silage looks drab and smells foul. Grass is the most commonly ensiled crop, but other green crops such as maize, lucerne and pea haulm are often used.

Unlike haymaking, fine weather is not essential for making silage. The crop is cut with a forage harvester or a mower. It is either cut and carted straight away or cut and left to wilt for a time, so as to reduce its moisture content.

The silos found on farms today are basically of two types – clamps and towers. Clamps are simply heaps of silage on the ground, usually draped with plastic sheeting to cover and seal them. A wall of concrete or sleepers often surrounds the heap – this reduces silage losses at the sides. Walled clamps are ideally suited to the self-feeding of silage by animals. Below-ground clamps are normally referred to as pits.

Grass drying This is the best way of preserving grass, but it is also the most expensive. The popularity of the process has declined in recent years owing to the sharp increase in energy prices. Nowadays, it is generally done on a factory scale and

Forage harvester

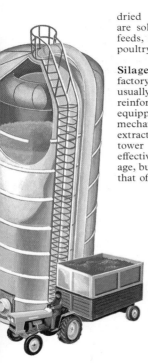

dried grass and lucerne are sold as supplementary feeds, usually for pigs and poultry.

Silage towers These are factory-made, airtight silos, usually of coated steel or reinforced concrete and equipped with various mechanisms for filling and extraction. An air-tight tower is easily the most effective way of making silage, but the cost far exceeds that of any clamp.

Cereals

Nearly 4 per cent of the world's entire land surface is devoted to the production of cereals and if all the cereals harvested were distributed equally throughout the world's population each person would receive about 250 kg per year. By world acreage, wheat is the most widely grown, then rice (which is less important in Europe), then maize (which is dealt with under Forage Crops, p. 230), then barley, oats and rye.

Before 2000 BC wheat was being cultivated in Britain on the higher land above the marshes and forests of the West Country. By 1750 more than half of the population ate wheat bread, which is more palatable and digestible than the coarser bread from other cereals. Wheat has been used here as a typical cereal to illustrate the various processes of cultivation.

Wheat

Oats

Barley

Rye

Wheat Wheat used to be grown mainly on heavy clay land but cultivation has now spread to the medium and lighter soils. For high yields the soil should not be too acidic, about pH 6 is optimal on a sandy soil and pH 7.5 on a clay soil.

Most wheat is produced from the eastern side of the country where there is more sun and less rain. A wet autumn and winter delays sowing, a wet summer delays ripening and then the grain starts to sprout in the stook. The optimum temperature for wheat germination is 25–31°C (minimum 0.5°C) and the best temperature for growth is 29°C.

Traditionally, wheat followed beans or Red Clover and yielded well. Winter wheat can also be successful after ploughed-up grassland, but there can be pest problems such as frit fly, slugs, leatherjackets and wireworms. Now cereal tends to follow cereal, increasing the risk of infection by fungal diseases such as eye-spot and take all. Chemical ploughing is a precaution after a

Take all (left) and eye-spot (above), two common fungal diseases of wheat

cereal crop: in late September the preceding stubble crop is sprayed to kill off any vegetation and the next cereal crop direct drilled in October.

Winter wheat requires 'vernalization' – the plants must experience low temperature and growth during the winter before they change from purely vegetative growth (no ears) to reproductive growth (ears). Scandinavian winter wheats need a great deal of vernalizing, and so have to be planted in early autumn. British, French and German winter wheats need rather less, and Belgian and French still less and can be sown in January and February. Winter wheat outyields spring wheat and can be harvested much earlier, giving grain of good quality (9 tonnes per hectare as against 7 tonnes for spring wheat).

Ploughing to prepare the seed bed

Cultivation Ploughing should be done as long as possible before sowing, normally to a depth of 15 cm, but if there are large quantities of trash to be buried then the depth is increased. The type of seed bed varies with the field, but it should not be worked so fine that the surface seals over after heavy rain, nor so coarse that seed germination is low.

The amount of nitrogen fertilizer used depends upon the previous history of the field: too much nitrogen on a tall-stalked wheat will cause the stalk to collapse (lodging). However, a suitable chemical is available that can be sprayed on to reduce crop height.

The amount of seed required varies with latitude and time of sowing; broadly speak-

Sowing with a seed drill

ing the earlier sown (e.g. October for winter wheat) and the closer to the equator, the less needed (early in the south 125 kg/ha, late in the north 300 kg/ha). Most wheat seed sown is dressed with compounds to control fungal diseases and pests. Seed costs depend on quality: breeder's elite stock being dearest and the farmer's own seed cheapest. Lists of the best varieties to sow are published annually. Cereals are established better by drilling than by broadcasting (scattering on the surface) and harrowing. Depth of seed placement is critical to yield: seeds planted at 2.5 cm give double the yield of seeds planted at 10 cm. The drills can be force- or gravity-fed with a wide range of coulters (cast-iron shoes that cut a furrow).

Harvesting Time of harvest depends upon the year and the location, and can vary from July to October. Winter wheat is usually ready to be harvested about nine months after sowing, spring wheat after six months. Following a hot, dry summer grain can be combine-harvested in good condition, and if the moisture content does not exceed 14 per cent it can be stored in sacks or bulk without drying. When it used to be cut by a binder (shown above), wheat was harvested a week or two earlier than it is nowadays by a combine harvester. Other cereal crops are harvested at similar times, again depending on location and weather.

Grain storage The vast majority of grain is brought in at moistures ranging from 16 to 24 per cent. At 16 per cent it can be stored for some time in sacks but not in bulk. Grain up to 20 per cent can be kept for a short period in open sacks but is likely to overheat and begin to go mouldy.

Types of drying Grain in sacks can be dried on a raised platform through which hot air is blown. Where tanker combines are used and the grain is handled in bulk continuous flow driers are better, with the grain passing slowly over a heated area. If the air temperature is too hot subsequent germination is reduced. Drying can also be carried out slowly in the storage bin but the simplest system is on-the-floor drying in which warm air is blown from underneath the grain.

Grain is commonly kept in a storage bin, but it must first be dried

Wheat products All the wheat grown in Britain is now bread wheat, but more is used for animal feed (48 per cent) than for human food (44 per cent). Of the latter 70 per cent is for bread, the remainder for cakes, biscuits and household flour. Only 20 per cent of the wheat flour used for bread is homegrown, the rest is strong wheat imported from Canada and the USA. Strength relates to the flour's ability to produce large loaves with good crumb texture. 'Weak' wheats, mostly European, make only small loaves with a coarse open structure. The strong wheats are the best for milling because the white endosperm (the flour) is more easily separated from the brown seed coat or bran. The bran can be further divided into the coarse, brown outside of the grain and the sharps and middlings, the pale internal layers of the skin. Wheat germ is the embryo. Three-quarters of the grain is flour, one fifth is bran, middlings and sharps; these are important animal feeds. A good bread wheat produces elastic dough whereas biscuit making requires plastic or sticky dough.

Grinding grain into flour is a combination of shearing, scraping and crushing. White flour has had the offals (bran, sharps, middlings) removed, whole-meal flour contains all the grain, whereas brown flour contains some of the offals.

Rye Rye is of more recent origin than wheat or barley. Pliny mentions it as a cereal new to the Romans. In the eighteenth century over a million acres of rye were grown in Britain and rye bread was eaten by about one-seventh of the population. This acreage has now declined. Rye has an extensive root system and is more drought tolerant than wheat, barley or oats; it is found on light, sandy soils unsuited to the others. Normally sown in mid-October, it is the best winter cereal for spring grazing by sheep or cattle because it produces the most ground cover and the highest weight of forage.

Rye protein contains no glutenin so rye dough is less elastic than wheat. It makes a heavy dense bread, popular in Scandinavia and Eastern Europe. Rye grain is used to make rye whisky or, in Holland, gin, as a filler for sauces, soups and custard powders and in animal feeds. Rye starch is a main ingredient of adhesives.

Barley Barley was probably the first plant to be cultivated and was believed to be the gift of Ceres. It has a very wide geographical range, growing from inside the Arctic Circle to tropical India, where it is grown at heights over 4500 m. Barley was grown in Britain during the Iron Age or earlier, first, it is believed, on the upland chalk soils as it is better suited to well-drained, light-to-medium soils with a low acidity. It is now grown chiefly in the drier eastern and southern regions of Britain with moderate rainfall and plenty of sunshine.

Winter barley is normally sown in October and November following early-harvested sugar beet, a short-term grass ley or after a spring cereal. Like winter wheat, it needs vernalization and the requirements of different varieties of winter barley vary considerably. Spring barley forms 80 per cent of that sown in Britain. It may be drilled from January to April.

As with wheat, chemical weed control is a usual spring operation on most barley crops and gaps in sowing, or tramlines, are often left to allow passage for the wheels of machinery in subsequent cultivations.

Once the barley crop starts to ripen there is a constant movement of plant foods from the leaves to the seed. When this movement ceases the crop is ripe. The leaves, stem and ears change to pale yellow and the ear bends over until the awns point to the ground. Winter barley ripens earlier than spring barley or winter wheat. The time of harvest varies from region to region. The finished crop is sold for malting, livestock feed, distilling, or for pearl barley or groats.

Using a spray boom to treat the green crop

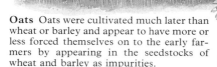

Oats Oats were cultivated much later than wheat or barley and appear to have more or less forced themselves on to the early farmers by appearing in the seedstocks of wheat and barley as impurities.

In Europe, wild oats were cultivated by cave dwellers in Switzerland before 1000 BC and the cultivated forms for human consumption arrived in Britain from Europe at a later date. The cultivated form of oats is different from that of the wild oat, which in Britain is regarded as one of the most troublesome weeds of arable land, but which is still grown for its grain and for green feed in south-western Asia.

The acreage of oats grown in Britain has declined to only about one-third of the area grown before the Second World War, being replaced mostly by barley. About 10 per cent of the oats grown is for human consumption and 90 per cent for stock feed.

Oats can be grown in a wide range of soil types, including some of the poorest alkaline soils in north and west Scotland, and on the light sandy and manganese-deficient soils of the Western Isles. It is mostly grown in the cooler, more humid climate of west and north Britain, where growth is relatively slow and the grains have plenty of time to grow plump.

Oats are usually sown in February to April, but winter oats can be drilled in early October. They cannot be grown too frequently on the same land because of susceptibility to cereal root eelworm.

Harvesting is often a week or two before full ripeness, partly because of the improved feeding value of the straw at that stage and partly because the grain is much more likely to be shed when fully ripe.

The wild oat is still grown for its grain in some parts of the world

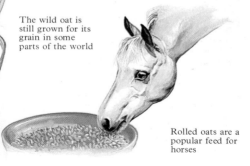

Rolled oats are a popular feed for horses

Like barley grain, the husk is tough, fibrous and quite inedible by humans. It therefore has to be removed by a special shelling process. Millers prefer oats with the highest kernel (groat) content and the least husk. Oat grains are higher in fibre and oil than barley and the protein in oats does not form gluten when mixed with water and cannot therefore be used for making bread.

Humans eat oats as porridge, oat cakes and breakfast cereals, whilst by-products of oat husks are nylon, paper and board. Oat husks are also used for refining oils and resins and in the making of fungicides and preservatives.

Brewing Two important reactions take place between harvest and brewed beer: one is the changing of starch to sugar, the other the fermenting of sugar to alcohol using yeast. Directly after harvesting the grain is dried for safe storage and to check fungal and bacterial activity. It is then cleaned. The storage allows secondary ripening to take place so that there will be a high rate of germination. After secondary ripening the grain is steeped in water to accelerate germination. During this process the grain is modified by enzymes, starch changing to maltose (or malt sugar), with less modification being required for pale than for brown ale. The germinating grain is then dried again in a malt kiln and after this it is screened to remove the rootlets, which can be used for animal feed. The screened product is the malt and beer is brewed by fermenting the malt extract. Brewers' grains, which can be used for animal feed, are a by-product of the malt-mashing process, in which the malt is ground and steeped in hot water to encourage the starch to turn into sugar. Hops, for flavouring, are added at the next stage, followed by the liquor being seeded with yeast.

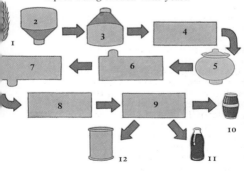

The brewing of beer: 1 Grains, 2 Malt bin, 3 Mash tub, 4 Filtering tank, 5 Brew kettle (hops added), 6 Starter tank (yeast added), 7 Fermenting tank, 8 Ageing tank, 9 Filtering tank, 10 Racking, 11 Bottling, 12 Canning

Straw Straw from the principal grain crops can be used for bedding, thatching or feeding. However, the cost of collecting and baling often proves uneconomic and it is simply burnt in the fields.

Thatching requires long quality straw. With the great increase in combine harvesting the most convenient crop is one with short straw, which is resistant to collapse, or lodging, particularly when large amounts of fertilizer are applied to increase yield. A combine tends to break the straw into small pieces, so thatching straw has to be cut by a binder.

Rye is the best thatching straw. It can grow up to 2.5 m tall, is almost solid in cross-section, is tough and wiry and can yield up to 4.5 tonnes per hectare. Rye straw is also used to make brown paper and packing material. It is much prized for bedding down horses but is least good for feeding.

214

Oat straw is best for feeding, one reason being that oats are often cut before they are ripe. As ripening continues the amount of soluble carbohydrates falls and that of in-digestible fibre increases, reducing the food value. Most straw can be used only as a maintenance ration for stock, and only ruminants can really deal with it. However, treatment with a solution of caustic soda increases digestibility. Oat straw can yield about 3 tonnes to the hectare, long-strawed wheat 4 tonnes.

Pea and bean straws are richer in protein and calcium than cereal straws, but their stems are less digestible, particularly bean stems which can get very woody.

Sorghum Sorghum, grown in similar areas to maize but more resistant to heat and drought, is of four types: grain sorghums (with white, yellow, red, brown or mixed grain), forage sorghums (which have sweet stalks), broom corn (whose brushes are used for making brooms), and grass sorghum (which is used for pasture and hay).

Sorghum is planted in late spring, when the soil is warm, in weed-free land – young sorghum is slow-growing and easily smothered by weeds. It can grow very tall, making it difficult to harvest, but there are dwarf varieties. Another problem is that the heads have to be cut from the stalks. Harvest is normally in late autumn or early winter when both grain and plant are dry. The small, hard kernels need grinding for livestock feed, except for consumption by sheep and poultry. Sorghum grain is also used for tapioca and in adhesives.

Rice Rice, a member of the grass family with a starchy kernel, is a high-energy food. The protein is in the husk and seed; when these are removed food value is lost. It needs relatively warm temperatures right through the growing season, water for irrigation and level, water-holding soils with adequate surface drainage. It is grown in southern Europe in Italy and France.

Rice is seeded in April to June. When plants reach 5–20 cm high the fields are flooded to 3–5 cm and as the plants grow more water is added. The land is submerged for 60–90 days and when the rice panicles begin to turn down the water is drained off. Harvest is about a fortnight later.

Rice is mainly used for human food. It is dried, cleaned, hulled and milled, the bran and seed being partly or wholly removed by scouring. Brown rice is whole kernels from which hulls have been removed. Polished rice has sugar syrup and talc added to produce a brighter shine.

Root, Forage and Break Crops

Although they are more expensive to grow than most other crops, roots (such as potatoes, sugar beet and carrots) are profitable cash crops. They make a useful break (or change) crop in an arable rotation, giving the land a rest from cereal-growing and breaking up and clearing the ground, Roots are now grown mainly for human consumption, whereas very large amounts used to be grown for feeding livestock in winter. This role is now mainly fulfilled by forage crops, which are cheaper to grow and require less labour.

Forage crops are grown as a source of greenstuff for livestock – either for grazing, for cutting and feeding elsewhere, or for making into silage. Important forage crops include kale, forage rape, maize and forage rye. Kale especially is a useful source of grazing for cattle in the autumn and winter, a time when there is no effective growth of grass.

In addition to root crops, peas, beans and oilseed rape are important arable break crops. Peas and beans are legumes, and so add nitrogen to the soil which can be utilized by the next crop in the rotation. Peas and beans are nowadays grown chiefly for human consumption rather than for feeding to livestock as a protein source.

Potatoes Potatoes are the most popular vegetable in Britain – we eat on average just under 2 kg per person per week. Those unfit for human consumption are fed to farm livestock. There are over 200 varieties of potato, about 50 of which are grown commercially. British farmers produce 6 million tonnes of potatoes in a typical year. There are three types of potato growing: early, maincrop and seed. Early (or new) potatoes are potatoes which are harvested while they are immature, and eaten straight away. Their production is concentrated in areas of the country which have a suitably mild climate. Pentland Javelin, Home Guard, Maris Bard and Maris Peer are some of the important early varieties. Maincrop (or ware) potatoes provide the bulk of the potatoes we eat, and the crop can be grown throughout the country. They are planted later, take longer to grow, and are only harvested when mature. Important maincrop varieties include Desiree, King Edward, Maris Piper and Pentland Crown. Seed potatoes are potatoes grown to provide seed for the next year's crop. Most of these are grown in special districts where the cool climate and altitude are particularly suited to producing vigorous seed potatoes.

Potatoes grow well in almost any well-prepared soil which is made ready in the autumn for the planting of earlies in January and February and maincrop potatoes in March and April. A few weeks before planting, seed potatoes are usually moved to a light, warm building called a chitting house, to encourage them to sprout (or chit). Seed tubers are planted by a machine, called a planter, in rows about 90 cm apart with some 28 cm between each potato, often less with early varieties. Modern planters are fully automatic and set two or four rows at a time, ridging up the

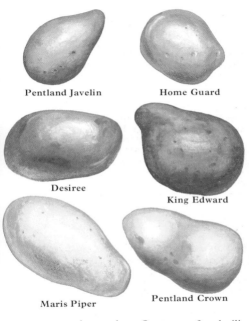

Pentland Javelin

Home Guard

Desiree

King Edward

Maris Piper

Pentland Crown

rows as they go along. One tonne of seed will produce about 10 tonnes of potatoes.

The growing crop must be protected from weeds, diseases and pests. The most common disease is blight and one of the commonest pests is the potato cyst eelworm. The Colorado Beetle, a very serious pest in most European countries, has been prevented from establishing itself in Britain. It does tremendous damage to the crop, completely stripping the haulm of its leaves.

Early potatoes are harvested from June onwards and most of the main crop in September and October. Potatoes are lifted with either a spinner, elevator digger or complete harvester. Hand picking is necessary after the spinner and digger, and the crop then carted away. The harvester does the complete job. Earlies are normally lifted with the haulm intact. A couple of weeks before lifting maincrop potatoes the haulms are often burnt off using chemicals, leaving the tubers in the ground so that their skins become firm.

Maincrop potatoes are commonly stored in pallet boxes in special buildings. Sometimes they are stored loose, surrounded by straw, in a cool, dark shed.

Colorado Beetle
A serious pest of potato crops in Europe but so far kept out of Britain

Sowing seed potatoes
with a planter

A potato clamp, now
rarely seen in Britain

Sugar beet Until recently, nearly all sugar consumed in Britain was refined from imported sugar cane. Today, about half is produced from home-grown sugar beet. Sugar beet is grown only on contract to the British Sugar Corporation, mainly in the Midlands and eastern counties within easy access of BSC factories.

Sugar beet requires the right fertilizers and a deep, well-drained, clean soil. It will not tolerate acid soils and needs to be well-supplied with lime. It is susceptible to the sugar beet eelworm if grown too frequently on the same land. To avoid this, one year's sugar beet should be followed by two or three years of cereals.

The soil is prepared by deep-ploughing before Christmas and the crop sown by precision drilling in March or April. Sugar beet seed is usually pelleted, so that all seed is the same size. This allows the seed to be accurately spaced in the soil when drilling. The aim is to grow about 74 000 plants per hectare.

The date of harvest is determined by contract with BSC, in order to ensure a steady flow of beet to the factory from the last week in September until the factory closes towards the end of December. The beet is harvested mechanically, one to six rows at a time, depending upon the harvester. The beet is topped before lifting and the tops are fed to livestock. The pulp left after processing at the factory is also fed to livestock.

Automatic sugar beet harvesting

Other roots Mangels (also known as cattle beet and mangolds) are a high-yielding root crop, grown for feeding to sheep and cattle in winter. They grow best in warm, sunny conditions so are more common in southern Britain. English varieties, unlike most roots, grow with much of the bulbous part above the ground. They are sown in April or May and harvested in late October or November and, if they are to keep, must not be lifted until mature.

Turnips and swedes are widely grown for human and animal consumption, though their popularity has declined considerably. Turnips have hairy leaves and there are white- and yellow-fleshed varieties; swedes have smoother leaves, yellow flesh normally and purple, bronze or green skins. They prefer cooler, wetter conditions than mangels and are mainly grown in the north and northwest. Sowing is in May or June for autumn harvest, though some quick-growing varieties of turnip can be sown as late as August. Sheep often eat turnips in the ground. Manuring, cultivation and harvesting of mangels, swedes and turnips are all as for sugar beet; storage is indoors or in clamps.

Fodder beet, once widely fed to pigs, is occasionally grown for livestock, mainly cattle. It grows like sugar beet. Both tops and roots can be fed. Mangels and fodder

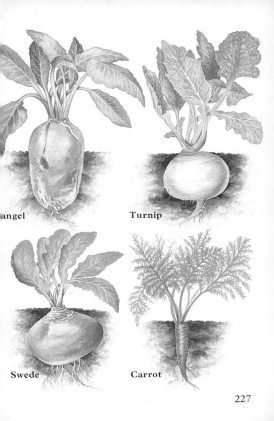

angel

Turnip

Swede

Carrot

227

beet are considered excellent for promoting milk production in lactating sheep.

Carrots are today a highly specialized crop grown for human consumption. Reject and surplus carrots are fed to livestock.

Forage crops Kale is the most important forage crop. There are two types: marrow stem and thousand head. Marrow stem kale is the most popular and gives the heaviest yield – 50–75 tonnes per hectare. It has a long, thick stem with leaves on top; both leaves and stem can be eaten. Thousand head kale has a shorter, thinner stem with a higher proportion of leaf. It is hardier than marrow stem but lower yielding. Traditionally, marrow stem kale is grown for autumn use and the more frost resistant thousand head kale for use late in the winter. Hybrid varieties such as Maris Kestrel are now available which combine advantages of both types. Kale is usually sown by drilling from March to June. The soil must be well limed and the crop given plenty of nitrogen fertilizer. Most of the kale grown is either grazed or zero-grazed by dairy cows.

Cabbage is mostly grown for human consumption, though still grown as livestock feed in some areas. Cattle cabbage is sown in late March to May and left in the field until needed. It is harvested by hand-cutting, and carted away to be fed to the

Thousand Head Kale

Cabbage

animals. The feeding value and characteristics of the crop are similar to those of kale. Three types of cabbage are grown for human consumption: spring, summer and autumn, and winter cabbage. Spring cabbage is grown either for cutting in December to April as unhearted spring greens or for cutting as hearted cabbage in April to early June. Summer or autumn cabbage is timed to follow spring cabbage. Winter cabbage is marketed from late December until hearted spring cabbage is available.

Forage rape is a quick growing forage which is ideal for catch cropping, especially in cereal stubbles. Its leaves are similar to those of swedes and turnips, but it does not have the same large root. It is mostly grown in the north and west of Britain and usually fed to sheep. The crop is sown by drilling or broadcasting between May and August.

Maize is grown to produce grain, sweetcorn or silage. The maize grown in Britain tends to be used as a forage crop; the whole plant above ground level – leaves, stem and cob – being made into silage. Maize requires sunny conditions and will not tolerate frosts, so it tends to be grown in the warmer parts of the country with a long growing season. The soil must be deep and well-prepared with good drainage. Seeds are drilled in April or May and the crop harvested, usually with a precision chop forage harvester, in September or October. One hectare yields 45–55 tonnes of maize silage. The mature crop can be left to frost before harvesting, as this increases its dry content. However, harvesting must be within ten days of frosting or deterioration will set in. Maize is very susceptible to bird damage: rooks in particular can wreck a crop by digging up the seeds and pulling out the young plants. Automatic scarers are often used to keep the birds away.

Forage rye is grown for grazing in early spring, particularly by intensive dairy herds and ewes and their lambs. It is well suited to cold districts. The varieties sown are special grazing varieties which differ from the ordinary cereal ones. Sowing is by drilling from August onwards.

Forage R

Maize plant and cob

An automatic bird scarer may be used to protect the young crop

231

Grazing forage kale

Other break crops Field beans, once an important source of protein for farm livestock, are now not widely grown. There are spring and winter varieties. Spring varieties include tick beans, an important cash crop often fed to racing pigeons, and horse beans, which like winter beans are suitable only for stock feeding. Winter beans, mostly grown in the southern half of Britain, are drilled in October, spring beans in February or March. Field beans prefer heavy, not too acid soils and need potash fertilizer to yield well. Average yields are about 3 tonnes per hectare. The crop is harvested by direct combining when the pods are black and the haulm shrivelled. Broad beans are grown for human consumption. The crop is harvested by cutting with a pea windrower and shelling with a mobile viner. Typical yields are about 3.5 tonnes per hectare. Green beans are grown on contract in the south and east of England for freezing and canning.

Peas are an important crop in the drier eastern counties of England. Most soils are suitable provided that they are not too acid. The bulk are grown for human consumption, mostly as vining peas or dry peas. Vining peas are cut green and grown on contract for canning, quick-freezing or dehydration. They are sown from January to

Harvesting field beans

May at a date set by the contractor, and harvested by cutting into windrows (swaths) by a windrower and shelling in the field using a viner. Average yields are about 5 tonnes per hectare. The green haulm is fed to livestock either fresh or after ensiling. Peas for harvesting dry are usually sown in early March. When the haulm turns yellow and the pods turn colour they are cut using a windrower and then combined. Crop yields are about 3 tonnes per hectare. Dry peas are used for processing, as seeds, or for sale as dried peas.

Oilseed rape is a popular arable break crop grown on contract to produce oil-rich seeds. It does not do well on acid soils. It is different from forage rape and the two are not interchangeable. There are spring and winter varieties. Winter rape is drilled in August and harvested the following July, spring rape in March and harvested in September. Winter varieties give higher yields of seed: typical are 2.5 tonnes per hectare for winter rape and 2 tonnes for spring rape. The crop is harvested either by swathing and combining or by direct combining. The oil is extracted by crushing the seeds and the meal left afterwards makes a valuable livestock feed known as oil cake.

Flax, a similar plant to linseed, is grown in parts of Europe. Linen is made from the fibres of its stem.

Broad Bean   Garden Pea   Oilseed Rape

Index

239